# Voice over
# Packet
# Networks

# Wiley Series on Communications Technology

Series Editors: Professor Barry Evans, University of Surrey and Professor David Hutchison, Lancaster University.

The *Wiley Series on Communications Technology* aims to present an up-to-date and in-depth picture of the converging areas of Telecommunications, Computing and Digital Broadcasting. Books published in the series will give a balanced view of Telecommuni-cations, Networking, Middleware and Applications. Providing details on the latest developments in Communica-tions, the series is aimed at engineers, managers, educators, researchers in academia and industry, and senior undergraduate and postgraduate students in computer science and electronic engineering.

# Voice over Packet Networks

**David J. Wright**
*University of Ottawa, Canada*

**JOHN WILEY & SONS, LTD**
Chichester · New York · Weinheim · Brisbane · Singapore · Toronto

*Other Wiley Editorial Offices*

John Wiley & Sons, Inc., 605 Third Avenue,
New York, NY 10158-0012, USA

WILEY-VCH Verlag GmbH
Pappelallee 3, D-69469 Weinheim, Germany

Jacaranda Wiley Ltd, 33 Park Road, Milton,
Queensland 4064, Australia

John Wiley & Sons (Canada) Ltd, 22 Worcester Road
Rexdale, Ontario, M9W 1L1, Canada

John Wiley & Sons (Asia) Pte Ltd, 2 Clementi Loop #02-01,
Jin Xing Distripark, Singapore 129809

***Library of Congress Cataloging-in-Publication Data:***
Wright, David J.
  Voice over packet networks/David J. Wright
    p. cm
  Includes bibliographical references and index.
  ISBN 0-471-49516-6 (alk. paper)
  1. Internet telephony.   2. Packet switching (Data transmission).   I. Title

  TK5105.8865.W75 2001
  621.382'12--dc21                                          00–068509

***British Library Cataloguing in Publication Data***

A catalogue record for this book is available from the British Library

ISBN 0471 49516 6

Typeset in 10/12pt Times by Deerpark Publishing Services Ltd, Shannon, Ireland.
Printed and bound in Great Britain by Bookcraft (Bath) Ltd, Midsomer Norton.
This book is printed on acid-free paper responsibly manufactured from sustainable forestry, in which at least two trees are planted for each one used for paper production.

To Mina, Leila and Ramin

# Summary Contents

# Contents

# Preface

This book presents the state of the art in packet voice. It is designed for readers who are studying and working in the forefront of the fast moving telecommunications industry. As such it is designed for readers who need information fast. The book streamlines its presentation of the subject matter by using the following techniques:

- It is a short book. The phraseology is very concise. There is no fluff.
- The format of the text uses extensive headings, subheadings and lists of points so that the reader can rapidly scan for the information that is relevant to their needs.
- Diagrams are used extensively, since a picture often adds clarity to a concise textual description and since telecommunications is a visual subject. There is no point in understanding a technology unless it is related to a network diagram.
- The style of writing is changed throughout the text from formal to informal and back to formal again, in order to provide the reader with some variety.

This book combines business and technology aspects of packet voice. It is primarily a technology book with about 20% of the content related to business issues. This is a balance that the author has found to be required by professionals in an industry which is primarily technology-based and where an in-depth appreciation of business issues must be founded on an understanding of the technology.

The book is designed to provide a bridge between circuit-based and packet-based telephony.

- For readers with a background in the circuit switched telephone network, it provides an introduction to each of the packet technologies in addition to describing how those technologies can be used to transport voice.
- For readers with a background in packet-based data communications, it provides an introduction to voice coding, features and QoS requirements as well as showing how voice can be transported over packet networks.

The book is organized in three parts:

- Part 1 describes issues related to packet transport of voice that are independent of the transport technology.
- Part 2 provides a chapter on each of the transport technologies: IP, ATM and frame relay and shows how voice is transported in each case. A final chapter provides an executive summary comparing the strengths and weaknesses of each technology.
- Part 3 uses the information from Parts 1 and 2 to describe how certain technologies can be combined to provide voice over broadband packet access networks.

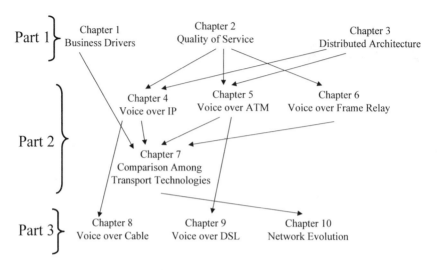

**Fig. 1.** Voice over packet networks: chapter dependencies

Many readers will come to this book already possessing a knowledge of some of its content. Fig. 1 presents the dependencies among the chapters, which readers can use to home in on the material they need.

# Introduction

This book provides information on how voice is transported over packet networks: a topic which bridges between the circuit networks from which the voice traffic is being migrated and the packet networks onto which it is being migrated. Interworking between circuit and packet transport is the name of the game while that migration is in progress. This book is therefore designed for readers with a background in the circuit world as well as for readers with a background in the packet world. The purpose of this introduction is to provide a primer to bring each reader up to speed on whichever technology they may not be familiar with. This material is very introductory, covering the basic concepts, terminology and acronyms and may well be skipped by many readers.

## 1. Introduction to The Circuit Network

The PSTN, public switched telephone network, transports voice and switches it in minimum bandwidths of 64 Kbps. Customers use a range of technologies to access the PSTN:

- *Analog twisted pair copper phone lines.* These lines run typically from residential customers to the network and carry an electrical signal representing the analog audio waveform in the region 300 Hz to 3.4 KHz.
- *Integrated services digital network (ISDN).* These digital lines provide two options:

  Basic rate interface (BRI) provides two user channels at 64 Kbps and one signaling channel at 16 Kbps
  Primary rate interface (PRI) provides 23 or 30 user channels at 64 Kbps and one signaling channel at 64 Kbps

- *T1/E1.* T1 is a North American standard with 24 channels at 64 Kbps. E1 is a European standard with 30 channels at 64 Kbps, plus signalling and management channels.

We now describe a day in the life of a telephone call in order to introduce the reader to all the components of the PSTN and how they work together.

A customer (called the calling party) picks up the phone which results in an "off-hook" message being sent to the network equipment, which responds by generating a dial tone for the customer to hear. This confirms that the network is ready to accept a call. The customer dials a number (called dialed digits) in order to place a call to a called party. The digits are sent down the access line from the customer to the network in one of a variety of formats:

- *Dual tone multifrequency (DTMF).* This is used on analog access lines for the dialed digits themselves.
- *Channel associated signaling (CAS).* This uses some of the least significant bits in the voice channel to transport codes corresponding to the dialed digits.

- *Common channel signaling (CCS)*. This uses a single channel to carry the codes corresponding to the dialed digits of a number of calls.

The dialed information is converted into signaling messages by the edge network equipment, also known as the ingress node or Class 5 switch. The signaling messages are interpreted by the switch in order to figure out the next hop on the path to the final destination, and the switch reserves one time slot (8 bits) on the outgoing transmission facility. The outgoing transmission facility may be, for instance, a T1 (1.5 Mbps), T3 (45 Mbps) or STS-$n$ ($n \times 51.8$ Mbps) and it connects the ingress switch to the next switch. Whatever the transmission facility is, it is controlled by a master clock which ticks 8000 times/s. T1, T3 or STS-$n$ frames containing time slots are generated at each tick of the clock. The 8 reserved bits for our phone call repeated 8000 times/s correspond to 64 Kbps.

The signaling message is then passed into the SS7 network which transports it to the next switch downstream, where the next hop of the route is determined and the next time slot reserved. This process is repeated until we reach the destination edge switch, which is another Class 5 switch also known as the egress switch. The egress switch causes the phone to ring at the destination, and the ingress switch causes a ringing tone to be played to the calling party. When the called party picks up the phone (going off-hook) messages are generated backwards through the SS7 network to tell the ingress switch to stop playing the ringing tone and the two parties can start to speak to each other. This process is referred to as call establishment or call set-up.

Once the parties start to speak their voice is coded into 64 Kbps using the ITU-T standard G.711. This may be done in the customer premises equipment (ISDN phone or private branch exchange (PBX)) if the customer is connected to the network over a digital access line or it may be done in the edge switch if an analog access line is being used. The G.711 coder/decoder or codec produces 8 bits of code 8000 times/s which maps neatly onto the time slots that have been reserved for its transport.

The customer can activate a voice feature, such as "redial on busy", by dialing more digits after the call set-up is complete. This activates software in the Class 5 switches at either end of the connection which provides the feature to the customer. In the case of redial on busy, the egress Class 5 switch monitors the destination line. When the customer goes on-hook, it signals the ingress Class 5 switch over the SS7 network to apply a special ring to the calling party line.

Class 5 switches are the edge switches in the PSTN and implement millions of lines of software to provide features of this type plus special private network management features for business customers. Moving towards the core of the network we encounter Class 4 (or tandem) switches, so called because the voice traffic traverses them (or tandems through them) without accessing any software. Next come the Toll switches, which represent the entry points into the long distance network where billing software monitors the call holding time and charges the customer accordingly.

Customers can access databases in the SS7 network which, for instance, translate toll free numbers (e.g. 800 numbers in North America) into routable numbers, implement local number portability and check the customers credit for charging a call to a credit card or calling card. These databases are called service control points (SCPs), and these services are sometimes referred to as intelligent network services.

# 2. Introduction to Packet Networks

A packet is a structure in which (i) some data (in the payload) is attached to, (ii) a header (which provides routing information to get the packet to its destination), and optionally (iii) a trailer (which usually provides error control). The trailer is a code obtained by a mathematical formula applied to the payload and possibly also to the header, and is inserted in the trailer by a transmitting switch. The receiving switch recalculates the same mathematical formula. If the result matches up with the code in the trailer, it concludes that there is no error in the packet. If it does not match up, some error control schemes can figure out what the error is in the packet and correct it (error correction), others simply note that the packet contains an error so that it should probably be discarded (error detection). An example of an error control code is a cyclical redundancy check (CRC).

Each switch in a packet network has a *routing table* relating:

> destination address
> to
> output port number

There is a fundamental difference between a connectionless packet switched network (such as IP) and a connection oriented packet switched network (such as ATM and frame relay).

- *Connectionless.* The packet header in a connectionless network contains the address of the ultimate destination to which the packet should be sent, on the other side of the network. The intermediate switches figure out the output port for the packet from *routing tables* based on this ultimate destination address.
- *Connection oriented.* Messages are sent through the network from source to destination requesting a connection to be set up. These may be signaling messages from the customer or messages initiated by the network management system. Each switch looks in the messages at the final destination that the connection needs to reach and from its *routing tables* assigns an output port. A connection identifier is also assigned on that port, to distinguish this connection from all other connections on the link from that port to the input port of the next switch. The switch maintains a record of the connection in the form of an *input/output table* relating:

> input port number and input connection identifier
> to
> output port number and output connection identifier

A common analogy for a connection is a pipe across a network. Once the connection is established it consists of a sequence of connection identifiers, one on each link that the traffic needs to traverse. A message is sent back to the customer to tell them they can start sending, plus they are informed of the connection identifier to use on the first link (from the customer to the network). The customer puts this first connection identifier in the headers of their packets and sends them to the first switch. The switch determines what to do with the packets from its *input/output table*. In general the connection identifier on output from a switch is different from what it was on input, and the switch makes the necessary change to the header as the packet traverses the switch.

In the above description it is important to distinguish between the *routing table* and the *input/output* table. In a connectionless network the former is used but not the latter. In a connection oriented network the former is used at call set-up time and the latter is used once traffic starts flowing. This book uses the above definitions of these terms. Unfortunately there is no industry consensus about this terminology and the term *routing table* is often used to refer to both of these types of table.

In both connection oriented and connectionless networks, other ''fields'' in the packet header may be interpreted and manipulated by the switch. For instance, in IP version 4, there is a ''Type of Service'' field which can be used (in particular in DiffServ) to determine the priority the packet receives with regard to delay and loss as it traverses a switch. In ATM and frame relay there are fields consisting of a single bit that is set if the packet traversed a congested switch.

Networks change dynamically: transmission links fail, new switches are powered up, congestion builds up and dies down, and these changes can affect the routing of packets. Routing messages are continuously flowing through packet networks, informing switches of changes in network conditions. Standard routing protocols such as open shortest path first (OSPF) in IP and private network to network interface (PNNI) in ATM, are used to provide and analyze information about such changes. The net result when a switch analyses routing protocol messages is that it changes its routing tables.

- In a connectionless network, if a link becomes congested or fails, the routing tables may cause subsequent packets to avoid that link.
- In a connection oriented network, if a link becomes congested or fails, the routing tables may cause subsequent call set-up requests to avoid that link. If a link fails, existing connections traversing that link may be re-established using the new routing tables.

# Part 1

## Voice Packetization

Part 1 of this book describes those aspects of packet voice that are independent of the transport technology to be used. Part 2 address the issues which are specific to individual transport technologies: voice over IP, ATM and Frame Relay, respectively. The general principles described in Part 1 are divided into three chapters:

- business drivers, in which we segment the market for packet voice and delineate the business reasons behind its deployment;
- quality of service, in which we analyze how the service requirements for packet voice depend on the application and how network-based and customer-based procedures can combine to fulfil those requirements;
- distributed architecture, in which we investigate the reasons for distributing voice functionality among many servers and gateways in a packet network and describe the protocols for communicating among them.

# 1

# Business Drivers for Packet Voice

Voice traffic is the largest single type of telecommunications traffic in terms of both bandwidth and revenue generation. It constitutes approximately 50% of all traffic on both public and private networks, and from the public carrier perspective, it constitutes 80% of the revenue. The migration of voice from circuit transport to packet transport has major consequences for service providers and equipment vendors, particularly those who have a significant investment in circuit switching. The aim of this book is to describe the technology alternatives for packet transport of voice and the business drivers that are bringing about this monumental change in the telecommunications industry. This first chapter focuses on the business issues, but first let us clarify the breadth of what we will be including in the term 'packet voice'.

## 1.1 Packet Voice Defined

Voice is not just human speech. Voice is human speech plus dialled digits and also any fax and data that can be carried over a voice-band modem on an analogue twisted pair access line. Dialled digits are often used during the course of a telephone conversation, for instance for interaction with a voice messaging system, and are referred to as Dual Tone MultiFrequency (DTMF). Often the network cannot distinguish between the different types of traffic that it might receive from an analogue twisted pair access line and therefore it is important to include all of them in the concept of 'voice'.

When voice is digitized, the codec that is used may itself packetize the voice or the packetization may be performed later when the voice enters a packet network. The standard for voice coding used in the circuit switched network is G.711 [1], which generates a constant stream of 64 Kbits per second which is *not* packetized. If this code is to be transported over a packet network, such as Asynchronous Transfer Mode (ATM), it must be packetized at the edge of the ATM network using one of the ATM Adaptation Layers. A popular codec which *does* packetize human speech is G.729 [2]. These are examples of the many ways of coding voice which are described in detail in Section 2.5. Voice codecs are designed for human speech. Some of them such as G.711 may be used for coding voice-band data or fax. Others such as G.729 cannot, and the data/fax must be handled differently for packet transport as described in Section 3.1.

A major advantage of packetizing voice is silence suppression so that packets are not generated during pauses between phrases or silence from one person while another person is speaking. The bandwidth savings from packetizing human speech are approximately 60%. However, there are many other *business* advantages to be derived from packetizing voice which we now describe.

We examine the business drivers for packet voice:

- first according to which part of the network is being used to transport the packet voice;
- second from the viewpoint of which organization operates the network;
- third from the viewpoint of the application that the packet voice is used for.

## 1.2    Business Drivers for Packet Voice in Different Parts of the Network

### 1.2.1    Public Network Backbone

The first business driver for packet voice is traffic integration in the public network backbone (Figure 1.1). Most public carriers have ATM and/or Internet Protocol (IP) backbones which were initially built for their data traffic. Migration of voice to the same backbone gives them efficient network management. Instead of having to have one set of engineers and managers to run a data network and another group to manage a circuit switched network, they can just hire one group of people to manage a single packet backbone network with a single network management software system. So low-cost network management is the business reason for transporting all traffic including voice on the packet network backbone.

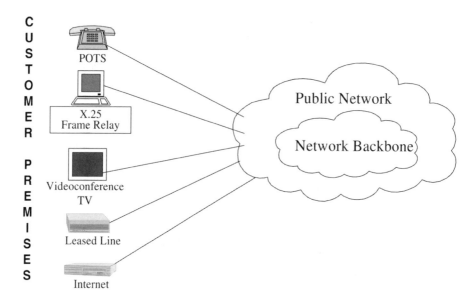

**Figure 1.1**    Traffic integration on public network backbone

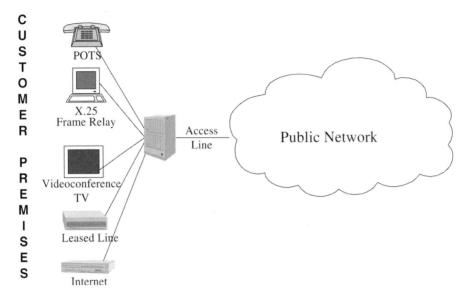

**Figure 1.2**   Traffic integration in access network

### 1.2.2   The Access Network

Packet transport can be used for voice, video, and data in the access network (Figure 1.2), just as it can in the network backbone, but the business case for using packet voice on an access line is different from the business case in the network backbone. First the business case in the access network is for the end-user organization instead of being for the carrier. Second, the business case is based on bandwidth saving as opposed to efficient network management. If a corporate customer has a packet access multiplexer then they can transport their voice, video, and data on a single access line instead of having several different access lines for different services, e.g. one access line for Frame Relay, another one for voice, another one for leased-line services, etc. So less access lines are required if we transport integrated traffic on a single packet-based access line. The business driver here is reduced access costs, whereas in the network backbone the business driver was reduced network management costs.

### 1.2.3   The PBX/LAN

The Private Branch Exchange (PBX) used to be a single piece of equipment typically in the basement of an office building, connected to the telephones of the people in that office building. Now the PBX is becoming distributed. The box in the basement is analogous to a mainframe computer, the functionality of which can be distributed to servers. The distribution of functionality of the PBX is shown in Figure 1.3 including the following functions:

- Gateway between analogue phones and packet-based Local Area Network (LAN).
- Gateway controller for setting up calls between gateways.
- Voice mail server.
- Feature server.

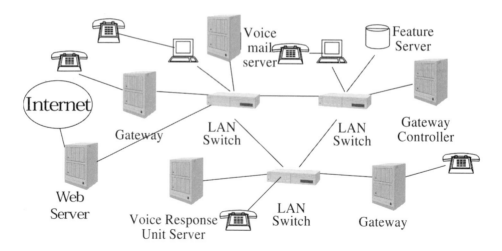

**Figure 1.3**   Traffic integration in the PBX/LAN

- Interactive Voice Response (IVR) server.
- Web server for Web-based call centre applications.

We need a transport technology to interconnect these servers, and the traffic which needs to be interconnected is the voice, which the PBX is mainly serving, plus the signalling traffic and Web page traffic, which is data, plus emerging multimedia conferencing applications and unified messaging. Packet transport is therefor a natural choice for the technology to use. The servers and gateways can be connected to each other using LAN switches.

So the business driver for packet voice in the customer premises network is the need to transport voice over the same network that is used to transport Web pages, signalling, multimedia conferencing and unified messaging.

The business case for the distributed architecture includes the ability of vendors to specialize in one function, the ability of customers to build a multivendor LAN/PBX, and the reliability that comes from having redundant servers and gateways.

### 1.2.4   The Desktop

The main desktop technologies are IP and ethernet. Since both are packet-based, conventional telephony uses a separate network and wiring to connect the desktop telephone to a PBX or Centrex service. The business driver for integrating telephony with data traffic at the desktop is first, reduced PBX interface costs and/or reduced Centrex charges and second, the ability to use networked multimedia applications. The investment in ethernet interface cards can be leveraged to carry voice and video as well as data traffic.

## 1.3   Business Drivers for Packet Voice for Network Operators

### 1.3.1   The New Competitive Public Carrier

The new public carriers competing with the incumbent long distance telephone companies are often building their networks in a greenfield situation. They do not have an installed base of circuit switched equipment, and can install a single transport technology suited to all traffic types. Since they are typically providing virtual private network services to end-user organizations, they purchase packet switching equipment because they have a variety of traffic going over those networks – some voice, some data and maybe some video as well. So they have got a good business case for packet voice. Because it is a greenfield situation it is natural to transport the voice over packet transport since that is what is used for data traffic. The net result is reduced operating costs because they have only a single network to manage.

### 1.3.2   The Cellular Company

Digital cellular is a rapidly growing industry. Voice is transported in digital format over the wireless portion of the cellular network using standards which already packetize the voice and compress it so that when it comes to the wired portion of the network, the voice is *already* packetized (Figure 1.4). It therefore makes sense to use a packet technology in the wired portion of the network. Packet transport can also handle the data traffic which the cellular company needs to transport including their own Operations Administration and Maintenance (OAM) traffic, which needs to be sent over the wired portion of the network. In addition, wireless data applications such as short messaging and wireless Internet access can be transported over the wired packet network. So the first part of the business case for packet voice in the cellular network is the need to find a common technology to transport cellular company internal data traffic, customer data traffic plus packetized voice; packet transport is a natural choice. The second part of the business case is that because of the rapid growth of digital cellular traffic the cellular company is often working in a greenfield situation either because it is a new company building a new network, or because it is an incumbent carrier expanding its network because of the rapid market expansion. Therefore, much of the deployment is in a greenfield situation where the business case for packet transport is easier to justify as in Section 1.3.1.

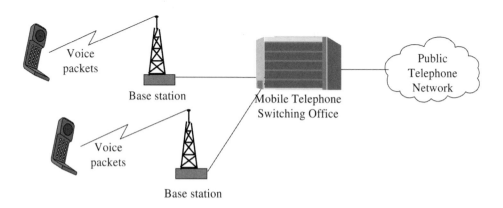

**Figure 1.4**   Cellular trunking

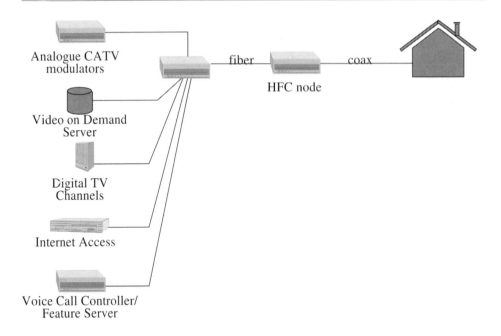

**Figure 1.5**   Cable company: broad service mix

### 1.3.3   The Cable Company

Cable companies were not initially in the business of telephony. Their bread and butter business was the distribution of entertainment video, followed by entering the market for Internet access. Digitizing their networks and making them bidirectional was necessitated in order to transport additional digital TV channels, to improve video quality, to carry upstream requests for video on demand and to provide high speed upstream and down-stream Internet access (Figure 1.5). The associated infrastructure cost is high, and the cable company needs to offset the capital cost of that infrastructure over as many services as possible. Voice is an additional service the cable company can add to their mix, but it is not a popular one with cable companies because of the complexity of the billing. The other services offered by cable companies are billed as a flat rate monthly fee, but telephony billing is required per call. Additional billing systems are therefore required for voice. However, it is another revenue stream which the cable company can use to offset the capital cost of upgrading the infrastructure. So for cable companies that choose to offer telephony, packet transport is a natural choice because of the need to find a common technology for the existing data, and digital video plus the voice.

### 1.3.4   Corporate Network

Corporate communications managers need to trade off users demands for bandwidth and quality of service (QoS) against the requirement for a cost efficient network. There was originally a reluctance to migrate voice onto the packet-based data network, because of concerns about QoS. However, for much intra-corporate communications a lower voice quality is acceptable than in the public network, and therefore packet transport was used for voice on corporate networks ahead of it being used in public networks. Cost reductions occur for two reasons:

- first only one integrated network needs to be operated and managed instead of having separate voice and data networks;
- second the packetization of voice saves about 60% of the bandwidth as a result of eliminating the silence periods.

The business case for packet voice in the corporate network is primarily based on these two sources of cost savings.

## 1.4   Business Drivers for Packet Voice Due to New Applications and Features

When voice is packetized it is usually integrated with data traffic for transport. This integration can result in new features being added to a telephone call. Customers regard features as being as important as the telephone call itself. On the circuit-based Public Switched Telephone Network (PSTN), customers pay as much for call forwarding as they do for the call that is forwarded! Some new features are specific to Internet-based calls, which we discuss first, and then we move on to features which are independent of transport technology.

### 1.4.1   Internet-Related Applications and Features

When voice is packetized all the way to the desktop, it opens up vast possibilities for new applications and features to be added onto Plain Old Telephony Service (POTS). This is particularly true when the packetization is done over the IP, since many Web-related functions can be added to telephony. Indeed many advocates of Internet Telephony to the desktop regard these additional functions as being as important as telephony itself, so that they provide an important element in the business case for Internet Telephony. Sections 4.6 and 4.7 of this book provide detailed examples of the additional functions that can be associated with telephony and here we give a brief summary.

- *Instant messaging*. If it is not possible to make a phone call because the destination is busy, a short message can be sent over the Internet which pops up on the screen of the called party.
- *Web integration*. When a customer is browsing a corporate Web site, they can initiate a phone call to speak with an agent by clicking a button on a Web page.
- *Collaboration*. During a telephone call to discuss a document, the document can be shared by the participants in the call. In 'view-only' mode document sharing is widely used for training purposes, in which an instructor gives a talk/slide show. In 'collaborate' mode it can be used for multiple authors to jointly edit a document.
- *Video telephony*. Adding video to a telephone conversation is particularly useful in situations in which the video displays the subject matter of the discussion, e.g. in equipment repair, health care and telelearning.
- *Voice and fax messaging*. Standards are in place for sending voice mail and fax as e-mail attachments over the Internet.

Additional services and features of these types constitute a strong element in the business case for IP telephony to the desktop. Although they could be provided over other packet technologies, we regard them as being specific to IP because IP is the dominant desktop packet technology. Some of these functions can be implemented in *end-user software*, so that the service provider does not expect them to generate a stream of revenue. However, they are of considerable benefit to the customer and can be regarded as swinging the pendulum in favour of IP as opposed to other packet technologies for transport of voice. In addition *network-based features* can be offered at a price, for instance, instant messages from the network to the user, e.g. to list current participants in a conference call.

### 1.4.2   Technology-Independent Applications and Features

In this section we describe two applications, Telephony Applications Programming Interface (TAPI) and Telephony Services Applications Programming Interface (TSAPI), which integrate circuit-based telephony and computer applications. TAPI and TSAPI are applications programming interfaces that enable a PBX and a computer to communicate. In the packet voice environment the functionality provided by TAPI and TSAPI can be implemented by software on a computer without the need to communicate to a PBX.

The first group of telephony functions are provided by TAPI (Figure 1.6). TAPI works at the desktop and it integrates the telephone with the desktop computer. TAPI was designed for outgoing and incoming *circuit switched* calls. In *outgoing* call establishment the user can have their address book on the computer, and click on the phone number they want to call. TAPI messages are sent to the PBX to set up the phone call. For *incoming* call management the call is routed according to the calling party number delivered from the network via the PBX. This is programmed into the software on the computer, and functions like voice messaging and IVR systems can be implemented on the desktop computer to handle incoming calls.

The advantage of implementing TAPI-like functions in a *packet voice* scenario is that we eliminate the need for the PBX. There is therefore no 'Interface' as such between the computer and the PBX. Instead we are using software on the computer to provide features similar to those described above for incoming and outgoing calls.

Note that with circuit-based TAPI we need a LAN connection for our data applications from the computer and a PBX connection for the voice connections. Providing TAPI-like features with packet voice simplifies the desktop wiring since we can have just one

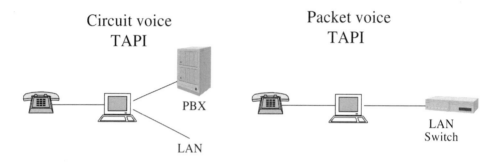

**Figure 1.6**   TAPI

Circuit Voice TSAPI          Packet voice TSAPI

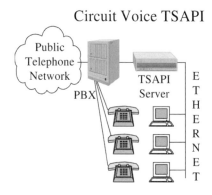

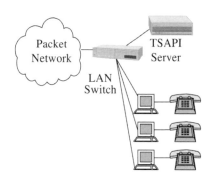

**Figure 1.7**   TSAPI

connection to the desktop for both the voice and data traffic instead of having two connections, one to the PBX and one to the LAN.

When TAPI functionality is implemented using IP, it corresponds with the IP desktop telephony situation described above. It could also be implemented using another packet technology, for instance, an ATM LAN.

The advantage of integrating the wiring becomes more pronounced in TSAPI, since we're not dealing with just one desktop, but instead with many desktops, in a call centre environment. Figure 1.7 illustrates agents' phones and computers on the same desktop. In a *circuit switched* environment, for incoming calls, TSAPI software in the server receives the calling party number from the PBX, pulls up the file of that customer onto the screen of an agent that is not busy, and at the same time it directs the PBX to send the voice call to the telephone of the same agent. When the agent answers the call, they've got the customer's file on their screen ready to rapidly do business with that customer. For outgoing calls, the server instructs the PBX to dial up a number of calls at the same time and whenever someone answers the phone, it directs the voice call to one of the agents and the profile of that particular client to the corresponding desktop computer. This is used for telemarketing applications.

The advantage of implementing TSAPI-like functions in a *packet voice* scenario is similar to the case of TAPI. We eliminate the need for the PBX and use software on the server to route calls and screen pops appropriately. All the voice, screen pops and associated signalling messages go over a LAN.

Packet-based TSAPI simplifies the wiring since only one LAN interface is required on each desktop instead of a LAN and a PBX connection.

When TSAPI functionality is implemented using IP, it is also referred to as a Web-based call centre which is described in more detail in Section 4.6. It could also be implemented using, for instance, an ATM LAN.

## 1.5  Summary

For any new technology to be accepted by the telecommunications industry, it must have a clear business benefit not just to a single player in the industry, but also to many other players as well. Packet voice has a very broad appeal. It is like an elephant around which the various industry participants gather (Figure 1.8). No single organization benefits from the whole elephant, but even that small part of the elephant that is immediately in front of each player gives them a reason to implement packet voice.

- The residential Internet user sees packet voice as a means of getting free telephone calls to anywhere in the world.
- The corporate telecommunications manager sees packet voice as a means of integrating voice and data traffic onto a single network and hence saving operating and management costs.
- The Internet Service Provider (ISP) can offer telephony as a service and hence generate a new revenue stream.
- Cable companies that initially entered the telephony market used circuit switching technology. Packet switching offers greater efficiency and integration with Internet access traffic.
- The equipment vendor can sell new products and/or new interfaces on existing products.
- The telephone company with a large installed base of circuit switched equipment may initially regard packet voice as a threat instead of an opportunity. However, deregulation and associated competitive pressures are forcing established telephone companies

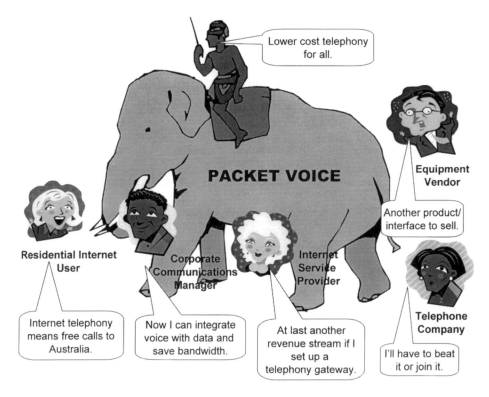

**Figure 1.8**   Packet voice

to reduce costs and packetization of existing traffic is one way of reducing transmission costs.

The first chapter of this book has presented:

- the diversity of applications for packet voice and their benefits to end users and service providers;
- the fact that there are business benefits of packet voice in all parts of the network, from the desktop, through the LAN and access network to the network backbone;
- the fact that every network operator can benefit from packet voice including cellular companies, cable companies, both new and established telephone companies and corporations.

In each of these cases the major elements in developing a business case for deploying packet transport for voice have been identified. Packetization of voice reduces bandwidth requirements and allows voice to be integrated with other traffic. In most cases this contributes to the business case by significantly reducing both the operating costs and the network management costs. However, each situation has its own specific determinants of the business case: the cellular company receives voice from the customers handset which is already packetized; the new telephone company does not have an installed base of circuit switched equipment; and the corporate desktop user needs to integrate telephony with a collaborative computing application. There are many reasons to transport voice over packet networks, and the vast volume of voice traffic that is being migrated to packet transport implies that this is one of the most fundamental changes to impact the telecommunications industry since analogue voice switches were replaced by digital circuit switches.

## References

1. ITU-T Recommendation G.711 (1988) Pulse Code Modulation (PCM) of Voice Frequencies.
2. ITU-T Recommendation G.729 (1996) Coding of Speech at 8 kbit/s Using Conjugate Structure Algebraic-Code-Excited Linear-Prediction (CS-ACELP).

# 2

# Quality of Service

## 2.1 Measures of Service Quality

### 2.1.1 Service Level Agreements

Service Level Agreements (SLAs) are contracts between service providers and their customers, which specify details of the service that is to be provided for voice and other traffic. If the service provider is unable to deliver the contracted service level, the SLA specifies a penalty, e.g. a percentage of the monthly bill that is refunded to the customer. The SLA includes the Quality of Service (QoS) plus a number of other aspects of the contract. QoS refers, in general, to technical measures of the service, and the other aspects included in the SLA are more business-related, including:

- problem resolution procedures;
- service availability, in terms of percentage up-time of the service;
- procedure for service changes;
- mean time to repair faults;
- call completion ratio, as a percentage of attempted calls that are successful;
- call set-up time (post dial delay);
- post pick-up delay (delay in obtaining dial tone);
- throughput in bits per second.

The QoS covers:

- accuracy: packet loss ratio and packet error ratio;
- latency: processing and telecommunications delay;
- jitter: delay variability;
- codec (coder/decoder) quality: the audio quality produced by the encoding and decoding of analogue voice into digital code.

The SLA parameters listed above can apply to individual connections through the network and/or to the total customer traffic. The business-related parameters such as service availability usually apply to all traffic, whereas the QoS measures will be different for data and voice traffic.

SLAs are important in a *competitive* service provider environment since they define exactly what services the customer will receive in a price-sensitive market. They are also important in a *co-operative* service provider environment where end-to-end connectivity among a customer's sites involves the networks of several different service providers. In this case the SLA defines which network operator is responsible for which aspects of the end-to-end service. Voice traffic needs to be handled carefully in such an environment because of its sensitivity to end-to-end delay, which is the cumulative effect of the delays introduced by each network operator en route.

### 2.1.2 QoS Parameters

The first three of the four QoS parameters, accuracy, latency and jitter, apply to all types of network traffic including data, voice and video. They are part of the design of any packet transport technology. Section 4.4 describes how ReSource reserVation Protocol (RSVP) and Differentiated Services (DiffServ) provide them to users of IP networks, Section 5.2.2 describes their implementation in ATM and Section 6.3 describes how they are incorporated in Frame Relay SLAs. But for voice, there is also a fourth element in the equation for providing service quality to the user, and that is the quality of the audio codec. Although the codec is evidently independent of the transport technology, it is often under the control of the network operator. For instance when a network operator sends circuit switched voice over a packet backbone, it passes through a gateway incorporating a voice codec together with mechanisms for encapsulating the voice code into transport packets.

Even though customers need to specify the QoS they receive from a network operator, it is important to realize that many aspects of QoS are under the control of the customer (Figure 2.1). For instance, delay is increased by customer premises equipment. Conversely, accuracy can be improved by customer premises equipment correcting or concealing errors introduced by the network.

Although accuracy can be improved, there is generally a trade-off between accuracy and latency. For instance, retransmission of errored packets improves accuracy, but introduces delay.

Methods of improving accuracy which introduce delay are unsuitable for real time interactive telephone conversations for which low latency is critical. However, they can be used in non-interactive streaming applications, such as voice messaging. Human interaction with a voice messaging system for sending or receiving voice mail does not need such a tight end-to-end delay as in the case of human-to-human interaction. The most impatient elements in a telecommunications network are the human beings, and two of

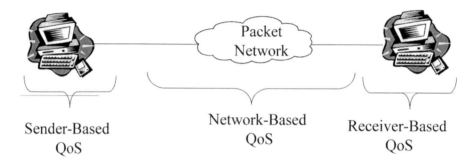

Sender-Based QoS

Network-Based QoS

Receiver-Based QoS

**Figure 2.1**   QoS: controlled by network and user

them, one at each end, creates the most demanding requirements on network design from the viewpoint of latency. Those latency requirements limit the range of accuracy improvement mechanisms that can be used.

The other situation which impacts accuracy is whether the communication is point-to-point or multicast. An example of a multicast situation is the delivery of a seminar using IP multicast. There are two types of participant.

- Some seminar participants may be passive, only requiring to listen to the seminar, without contributing comments or questions. They are essentially operating in streaming mode, receiving a stream of voice for which delay is relatively unimportant. They can therefore employ error correction mechanisms that improve accuracy and voice quality even though they also introduce delay.
- Other participants may need interactive discussion during the multicast. For them, latency is more important, and they may have to sacrifice accuracy in order to reduce delay.

The trade-off between accuracy and latency is therefore different not only for different voice applications, but also for different participants in a single application.

Network-based QoS and codec quality are parts of the same puzzle of how to produce the voice quality the user requires. They are therefore both discussed in the current chapter.

We now analyze each of the four aspects of QoS in turn.

## 2.2  Accuracy

Acceptable loss ratio and bit error ratio for voice packets depends on the type of voice that is being transported.

- Generally speaking, modem data is highly sensitive to information loss. Loss of even a short packet, such as an ATM cell, can result in higher level protocols retransmitting a much longer data unit of which that cell was part. Thus, a low loss ratio is necessary in order to reduce the amount of retransmitted traffic.
- Fax does not incorporate error recovery procedures such as retransmissions, however many errors go unnoticed to the human eye. Loss of a 50 byte packet in a page of a business document represents only about 0.1% of the total content.
- For human speech, information loss becomes a major concern if the speech is compressed. Loss of a 50 byte packet of 64 Kbps speech represents only about 6 ms of speech, which is less than a phoneme (a basic sound out of which human speech is composed), and therefore has a minimal effect on comprehension. Several such losses can result in a degradation in subjective voice quality. In the case of compressed speech, however, the effect is much more serious. Fifty bytes of 6.3 Kbps speech represents about 60 ms of voice, which can result in entire phonemes being missing. A low packet loss ratio is therefore important for the transport of compressed speech, particularly when there is generally not time for retransmissions. A common design objective for compressed speech such as G.723 and G.729 is a maximum of 1% packet loss ratio. For G.711 a higher packet loss ratio is acceptable. These loss ratios are for the loss of isolated packets. If packet losses occur in groups they have a much more serious effect on degrading the speech quality.

## 2.2.1  Sources of Information Loss

As voice packets traverse a network, there are three situations that they can encounter that may cause information loss.

- *Buffer overflow.* Any packet switch incorporates buffers that accumulate packets which are waiting to be switched or sent onto output ports. Due to the variability in the arrival rates of all the packets being handled by the switch, it may be necessary to discard some packets. Although this is commonly referred to as 'buffer overflow', it does not necessarily mean that the buffer is physically full. Buffer management mechanisms can also discard low priority packets when a buffer is only partially full, in order to leave space in the buffer for subsequent packets that may arrive with higher priority. Voice packets are usually given high priority, but not necessarily the highest priority, which may be reserved for mission critical data or high resolution video. Many buffer management mechanisms effectively operate by having virtual markers for each priority level on the buffer at different fill levels. When the buffer gets fuller than the marker, subsequent packets of the corresponding priority are discarded. Buffer overflow is by far the most common cause of packet loss in a packet switched network. It results in the loss of an entire packet, as opposed to the corruption of individual bits within a packet.
- *Laser malfunction.* Bit errors can be caused by temporary malfunction of the lasers and their control circuits that transmit light pulses along optical fibres. It is unusual for errors of this type to cause a single bit error. Instead a stream of consecutive bits is usually affected.
- *Network failure.* A network node such as a switch may fail due to electrical or optical component failure or due to power failure. A transmission line may fail by being physically cut by a back-hoe operator digging in the wrong place or by overhead lines between telegraph poles being brought down in a storm or a road accident. Many networks are designed to re-route all traffic onto back-up paths to cater for network failures of this type. For instance, Synchronous Optical NETwork (SONET)/Synchronous Digital Hierarchy (SDH) networks use a ring architecture and re-route traffic the opposite way around the ring in the case of a failure. This re-routing is done at the physical layer and is very fast, within 50 $\mu$s, causing negligible impact on voice quality. Virtual circuits in ATM networks can be set up with back-up virtual circuits that take an alternate path in the case of a failure. Switching traffic to the back-up path at the ATM layer takes longer than at the physical layer and a few 10s or 100s of milliseconds of traffic can be lost.

Figure 2.1 indicates that accuracy can be improved by employing sender-based or

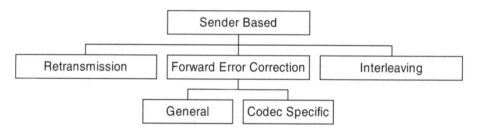

**Figure 2.2**   Recovering lost packets

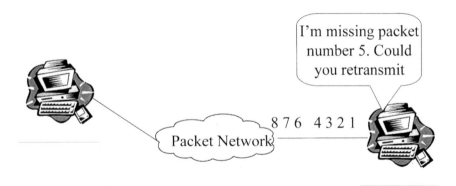

**Figure 2.3**   Sender-based: retransmission point to point

receiver-based mechanisms. A classification of sender-based mechanisms is shown in Figure 2.2. Receiver-based methods are shown in Figure 2.9 and are described in Section 2.2.3.

### 2.2.2    Sender-Based Control Mechanisms

#### 2.2.2.1    Retransmissions

Retransmission of errored or missing packets is the most widely used sender-based recovery method. It is used in Transmission Control Protocol (TCP) for data traffic. Figure 2.3 shows a sender of information on the left which has added sequence numbers to the packets transmitted: 1, 2, ..., 8. The receiver on the right can deduce that packet #5 is missing from the gap in the sequence numbers. Alternatively, the receiver may check out an error control sequence on the packets and find that packet #5 contains an error. Either way, the receiver requests a retransmission of packet #5 and acknowledges receipt of all other packets. A significant delay is introduced corresponding to the round-trip time for the request for retransmission to go from receiver to sender and the retransmitted packet to travel back. Retransmissions are therefore suited to streaming as opposed to interactive voice applications.

In a multicast environment, retransmissions operate in a different manner. Figure 2.4 represents a multicast in which any transmission by any of the four participants is multicast to each of the other participants. We take the situation of packets currently being transmitted by the source at the top left of the diagram. As in the unicast situation described above, packets are numbered by the source. Suppose recipient #1 is missing packet #5. It could be that the packet was lost on recipient #1's access line in which case they would likely be the only recipient to have packet #5 missing. Alternatively the packet may have been lost in the network backbone, thus affecting all downstream recipients. Since we do not know which of these two situations applies, we devise a mechanism that caters for both. Recipient #1 requests a retransmission of packet #5 from *anyone* in the multicast. Recipient #2 is also missing packet #5, but they receive the retransmission request from recipient #1, before they can send out their own retransmission request. There is no point in two retransmission requests for the same packet, so recipient #2 does nothing, and simply waits. In order to further reduce the chance of multiple retransmission requests, each retransmission request is delayed by a random amount of time

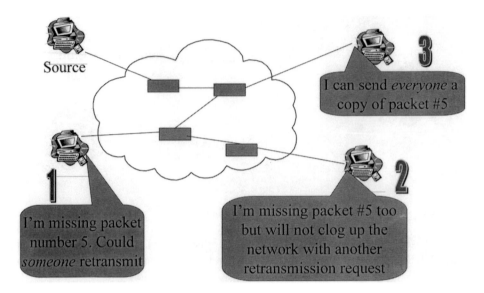

**Figure 2.4**   Sender-based: retransmission multipoint

whose mean is proportional to the distance of that receiver from the source. The first
participant who receives a retransmission request, and who has received the required
packet, #3 in Figure 2.4, retransmits the packet, thus ensuring that on average the retransmission will be from a receiver that is close to the point of failure. Other participants that
could have retransmitted do not do so when they see #3's retransmission. This mechanism
is suited to non-interactive participants in a low error rate environment, and can be used for
voice in applications such as training sessions and business presentations over packet
networks, particularly IP, for which multicast is well developed and efficient [1].

### 2.2.2.2   Forward Error Correction

Forward Error Correction (FEC) can be used for voice applications in two ways. The first,
illustrated in Figure 2.5, is a general purpose technique that is applied after the voice codec
has produced its voice packets. The second, illustrated in Figure 2.6, is used as part of the
voice codec itself.

   The general purpose method takes a block of voice packets and calculates a FEC code
that is sent as a separate packet or packets. An example of this is ATM Adaptation Layer
Type 1 (AAL1) (see Section 5.3.1), in which FEC is an option. If a voice packet is lost or is
errored by the time it reaches its destination, the FEC code may be used to recover the
original packet. This method introduces a delay dependent on the length of the block of
voice packets. A missing or errored packet anywhere within the block cannot be recovered
until the FEC packet is received, at the end of the block. For instance, in AAL1 124 ATM
cells are used in the block. This effectively limits the application of this method to non-
interactive, streaming voice. The method is particularly suited to high loss rate environ-
ments, where retransmissions would cause a significant increase in traffic in the network.
Residential coaxial cable networks often incorporate low quality cable connectors which

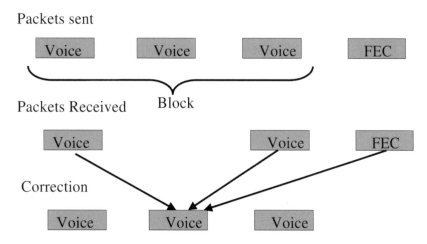

**Figure 2.5** Sender-based: FEC, general purpose

can cause high packet loss rates, and can use general purpose FEC for streaming audio and video traffic.

Codec-specific methods produce two versions of the voice code, the regular code and a more compressed code that can be used in case of errors in the regular code. Figure 2.6 illustrates that voice packet #3 contains a highly compressed version of the voice code in packet #2. If packet #2 is lost, it cannot be recovered in its original form, but the highly compressed version can be used instead. There is a slight degradation in speech quality at the receiver, but not as significant a degradation as if the packet were completely missing. RFC 2198 [7] specifies an implementation of this scheme that is suited to IP multicast. It does not introduce as much delay as the general purpose method, and is particularly appropriate for voice codecs producing long packet sizes, for which receiver-based interpolation schemes (Section 2.2.3) would result in a loss of intelligibility. Voice codecs used on the multicast backbone of the Internet (Mbone) use 20–80 ms voice packets with a default of 40 ms. Receiver-based interpolation methods can handle shorter missing pack-

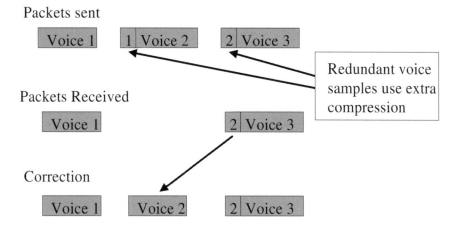

**Figure 2.6** Sender-based: FEC, codec-specific

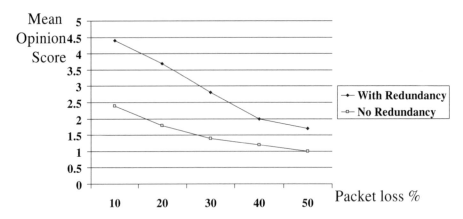

**Figure 2.7**   Benefit of redundancy

ets, but 40 ms packets can contain entire phonemes for which codec-specific FEC can be used to recover the phoneme. Figure 2.7 shows the benefit of including redundant low bit rate voice code in foreign language tutorials over the Mbone [2]. It is clear from the graph that this method is ineffective at high packet loss rates for which interleaving is more appropriate.

### 2.2.2.3   Interleaving

In situations of high loss rate, packet losses often occur in groups. This could be caused, for instance, by a congested buffer in a switch in the network. During congestion, many consecutive packets may be discarded, and when congestion abates, packet loss is more rare. FEC methods are often overwhelmed by groups of consecutive missing packets. The general purpose method has a limit as to how many packets in a block can be recovered and the codec-specific method can only correct a missing packet if the next packet arrives intact.

Interleaving is a method of spreading the effect of a group of consecutive errors over time. Figure 2.8 shows how a block of nine consecutive voice samples are interleaved before being transmitted. The transmission order is different from the original order of the voice samples, so that if a group of three consecutive samples is lost from the transmission sequence (#s 2, 5 and 8), they are not consecutive after being un-interleaved at the destination. This makes it easier for one of the above methods or for a receiver-based method to be used to recover from the losses.

Interleaving introduces delay dependent on the length of the block of voice samples or packets used, and is therefore suited to streaming voice applications.

### 2.2.3   Receiver-Based Control Mechanisms

Receiver-based methods generally introduce less delay that sender-based methods and are therefore more suited to interactive human voice conversations. They do not recover lost packets as accurately as sender-based methods. They essentially trade accuracy for speed,

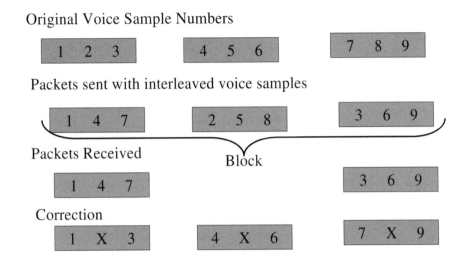

**Figure 2.8**   Sender-based: interleaving

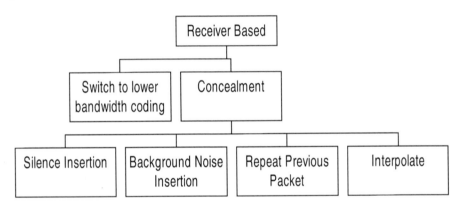

**Figure 2.9**   Recovering lost packets

which is the right compromise for interactive speech. The alternative types of receiver-based methods are shown in Figure 2.9 and are now described.

### 2.2.3.1   Switching to Lower Bandwidth Encoding

An IP multicast situation may be set up in which the receivers have a choice of different voice coding schemes. A receiver may start by requesting a high quality coding scheme, which consumes correspondingly high bandwidth. Subsequently, the receiver may notice a high loss rate in arriving packets, which may be due to network congestion. Voice quality is dropping as a result of missing packets and may be worse than would be obtained from a lower bandwidth codec. It is therefore in the destination's interest to switch to a lower bandwidth codec. This switch also helps to alleviate network congestion, so that subsequently, after congestion has abated, the destination can switch back to the original higher quality codec.

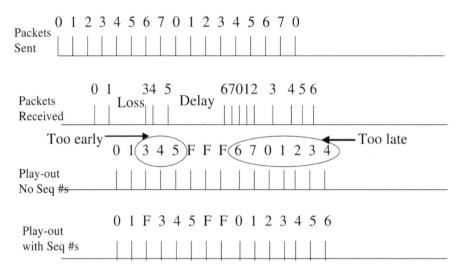

**Figure 2.10**  Benefit of sequence numbers

### 2.2.3.2  Concealment

Concealment methods hide the effect of packet loss and errors, instead of attempting to correct them. They are well suited to voice applications for which interactivity is more important than voice quality, and are incorporated in popular standard codecs such as G.723.1, G.728 and G.729.

Concealment methods are based on the use of sequence numbers to detect whether a packet or voice sample is missing or delayed. Figure 2.10 shows an original sequence of voice packets numbered modulo 8. Packet #2 is lost and packets #s 6 and 7 are delayed in the network. At the destination, playout of the voice packets is postponed to accommodate slight delay in the arrival of packets, but in Figure 2.10 the delay to packets #s 6 and 7 is too great for this mechanism to work on its own. Without the use of sequence numbers (third row of Figure 2.10), packets #s 6 and 7 are played out too late and dummy packets indicated by F, F, F must be inserted. The absence of sequence numbers also means the destination is unable to detect the fact that packet #2 is missing and packets #s 3, 4 and 5 are therefore played out too early. The use of sequence numbers (last row of Figure 2.10) enables a dummy packet to be inserted in place of packet #2. Dummy packets are also used instead of the late packets #s 6 and 7, so that the packets after #7 can now be played out on time. Ref. [3] indicates the number of sequence numbers to use in different circumstances. For instance, more sequence numbers are required if link capacity is low or if link utilization is high and eight or 16 sequence numbers are commonly used in commercial networks.

Figure 2.11 classifies five different concealment techniques according to how much processing is required to achieve a given improvement in voice quality.

The simplest method is to do nothing, i.e. to replace the missing packet by silence. This can be done with very short voice samples, up to a maximum of 5 ms, but thereafter it results in a degradation in voice quality. A significant improvement in voice quality at a minimal increase in processing can be achieved by inserting background noise instead of silence. This improves the three major measures of voice quality:

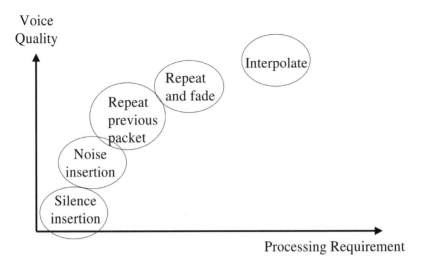

**Figure 2.11**  Concealment

- restoration of partly missing phonemes;
- intelligibility;
- subjective voice quality.

A further improvement in voice quality can be achieved by replacing a missing packet by the previous packet instead of replacing it by background noise. Again only a minimal amount of processing is required.

Global System for Mobile Communications (GSM) uses a method of repeating the previous packet for 20 ms and then fading the audio out over the next 320 ms. This requires significantly more computation for a marginal improvement in voice quality and is suited to situations such as GSM where the packet length is long.

Interpolation between the packets on either side of the missing one adds more to the computation requirement with only a very small improvement in voice quality and is not used very widely. Also it has the disadvantage of introducing a delay of one packet time.

### 2.2.4   Summary of Error Control for Voice

The receiver-based mechanisms introduce minimal delay and can therefore be used both for interactive voice and also for streaming voice applications. The sender-based mechanisms introduce significant delay and are therefore restricted to streaming voice applications, with the exception of codec-specific FEC which can also be used with interactive voice.

## 2.3   Latency

The majority of voice applications are interactive, for which latency is the primary QoS measure. Delay is measured end-to-end across the packet network from the point where the voice is coded at the source, across the packet network, to the point where it is decoded at the destination.

|              | G.723.1  | G.729 | G.729A |
|--------------|----------|-------|--------|
| **Kbps**     | 5.3/6.4  | 8     | 8      |
| **Frame size** | 30ms   | 10ms  | 10ms   |
| **Lookahead** | 7.5ms   | 5ms   | 5ms    |
| **DSP MIPS** | 16       | 20    | 10.5   |
| **RAM**      | 2200     | 3000  | 2000   |

**Figure 2.12**  Delay characteristics of codecs

### 2.3.1  Sources of Delay

Delay can be caused in a voice communication by many different factors.

- *Operating system delay for desktop applications.* Real-time operating systems are able to reduce the extent of delay by giving priority to voice applications, but many end users have non-real time operating systems which allocate time among all programs currently active in a round-robin fashion. Latency can therefore be reduced by closing down as many applications as possible except for the voice application.
- *Voice processing delay.* When Digital Signal Processor (DSP) chips are used to code voice, the processing time corresponds to the voice frame size: a few 10s of milliseconds (Figure 2.12). DSPs are widely used in voice gateways which are optimized for voice processing. Desktop computer sound cards, however, are not so optimized and can introduce a delay of 20–100 ms into a stream of voice.
- *Look-ahead processing.* When some codecs are coding a voice sample, they take into account information in the next voice sample, in order to give continuity to the audio waveform between the two samples. This introduces a look-ahead delay, which is given in Figure 2.12.
- *Packetization delay for voice packets.* If a single voice packet is to be transported in a packet, there is minimal packetization delay. However, if $n > 1$ voice packets are to be packed into a single transport packet, there is a packetization delay of $(n - 1)t$, where $t$ is the packet time. At the source, the first of the $n$ voice packets has to wait $(n - 1)t$ while the other packets are being coded. At the destination the last of the $n$ voice packets has to wait $(n - 1)t$ while the earlier packets are being played out.
- *Packetization delay for voice samples.* The same applies to voice samples of one byte per sample, as is the case in G.711 voice. Each byte is produced every 125 µs; therefore, the delay in putting $n$ bytes into a packet is $(n - 1) \times 125$ µs.
- *Modem delay.* In desktop applications using modem access, the DSPs in the modem can add to total delay to an extent of 20–40 ms per modem.

- *Ingress/egress delay.* The emission time on the ingress/egress access line into/out of the network can add to the total delay if the bandwidth of the access line is low. This is the time it takes between the first bit of the packet being put onto the access line and the last bit of the packet being put onto the access line. It is *not* the time taken to traverse the access line, which is included in network delay (below). Ingress/egress delay is significant for desktop multimedia applications using modem access. A 50 byte packet on a 33 Kbps access line has an emission time of $50 \times 8/33 = 12$ ms.
- *Network delay.* Although the network is often blamed for delay, particularly by novice users, it is only one factor in contributing to total delay. Network delay is highly variable. It includes propagation delay which depends on the end-to-end distance of the communication. For instance the speed of light in fibre is about 200 000 km/s, which introduces a delay of 15 ms on a 3000 km phone call. The other components in network delay are the switching delay, which is only a few microseconds, and the buffering delay in each switch, which depends on congestion.
- *Jitter buffer delay.* The buffering delay in network switches depends on other traffic volume and also on the priority of other traffic and introduces considerable variability into total delay (see Section 2.4 below). In order to absorb the variability in the delay between one packet and another, a jitter buffer is implemented at the destination. When packets arrive, they are not played out immediately, but are kept in a buffer. When a sufficient supply of packets is available in the buffer, so that it is unlikely that it will run dry of packets due to variability in arrival times, we start to play out the voice to the destination user. We therefore absorb network jitter by introducing delay at the destination.

The total of all these factors is the delay experienced by a user of a packet voice network, and is very different in a desktop-to-desktop situation from what it is in a gateway-to-gateway situation. We now give some ball-park figures for how delay accumulates in two situations.

Figure 2.13 indicates how the network delay can dominate the gateway-to-gateway situation. Total delay can therefore be significantly reduced by managing the network so that it introduces little delay for voice, e.g. by giving voice priority over other traffic types or by planning for plenty of spare bandwidth in the network.

Figure 2.14 indicates the very different case of a desktop-to-desktop situation where there is no single factor that dominates total delay. Instead, the network, sound cards, modems and operating system all contribute significantly to total delay. Moreover, the total delay in the desktop-to-desktop situation is much higher than in the gateway-to-gateway situation, reflecting the fact that gateways are optimized for voice processing whereas most desktop computers are not.

### 2.3.2   Controlling Delay

Figures 2.13 and 2.14 have shown how delay of the order of a few 100s of milliseconds can be introduced into a voice call. We now look at how important that is, and what can be done about it. There are two reasons why delay is important to voice.

To make a phone call over a network, the ITU standard G.114 specifies that the end-to-end one-way voice delay must be less than 150 ms. The reason why the ITU does not want the telephone company to introduce more delay than that is in order to maintain the flow in

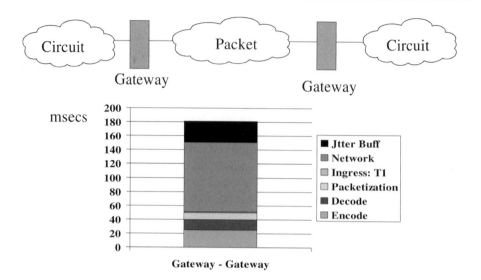

**Figure 2.13**   Total delay: gateway-to-gateway

human conversation. People who regularly use geostationary satellite links for phone calls can adapt to the delay of 250 ms introduced, but the average user finds that a delay above 150 ms restricts interactivity. Many users find a delay above 100 ms to be inconvenient and many networks are built with a 100 ms design objective. The upper limit of what is tolerated by the ITU is 150 ms.

Another reason why we're concerned with delay in a packet voice network is echo. Echo can be caused by an interface between part of the network which is four-wire and another part of the network which is two-wire (see the upper half of Figure 2.15). At this interface, we can get an echo where the speech, which was going from left to right, hits this

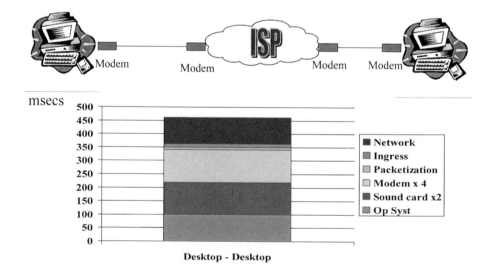

**Figure 2.14**   Total delay: dial-up access desktop-to-desktop

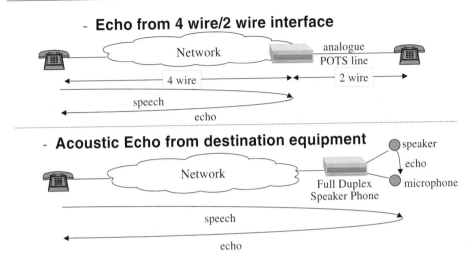

**Figure 2.15**   Sources of echo

four/two-wire interface and some of the speech gets echoed back to the person that was originally talking. If a speaker hears an echo of their own voice after a certain time interval, it sounds to them as if they are talking in a large hollow room. If the echo is returned to the speaker within a very short time interval it is not noticeable. Acceptable delay is within 15–50 ms depending on the loudness of the echo path [4]. This type of echo is caused typically by an interface between a digital network which is four-wire and an analogue POTS line which is two-wire. In order to reduce the effect of this echo, we can either reduce the amplitude of the echo, so that its so soft that the speaker doesn't notice it, or reduce the delay so that the echo comes back to the person so fast that it just sounds like their own voice rather than an echo of their own voice.

Another reason why we have to be concerned about echo is acoustic feedback, e.g. from a speaker phone to a microphone, as illustrated in the lower half of Figure 2.15. This echo is present even if we have four wires all the way through, from source to destination.

Echo can be removed by echo cancellation implemented in the DSP chips that perform the voice coding. The performance of echo cancellers is specified in G.168 [6] and each vendor implements their own design in order to achieve the required performance. Echo cancellers store the outgoing speech in a buffer and monitor the incoming speech to see whether it contains a component that matches up against the stored speech after a delay. If it does, that component in the incoming speech is cancelled out instead of being passed back to the user since it is an echo of what the user originally said. Echo cancellers can be tuned or can tune themselves to the echo delay on any particular connection. Each echo canceller has a limit as to the maximum delay of the echo that it can identify.

Echo is unimportant to human users if we've got regular telephones, no speaker phones, and we've got four wires end-to-end. Otherwise, echo is important to human users, but not to machines. Fax and modem data are insensitive to echo. Therefore, the voice gateway interface card needs sufficient intelligence to detect a fax or modem squeal and to bypass the echo canceller.

We therefore need to be concerned about delay in a packet voice network from the viewpoint of echo (15–50 ms allowed) more than from the viewpoint of maintaining the flow of human conversation (150 ms allowed).

It can be seen from Figures 2.13 and 2.14 that it is going to be extremely difficult to design a network and edge equipment which will transport voice with less than 15–50 ms end-to-end delay, whereas aiming for less than 150 ms delay so as to maintain the flow of human conversation is feasible. The primary mechanism for controlling delay by network design is to give high priority to voice traffic. This allows it to bypass lower priority traffic in switch buffers.

The solution to the problem of echo is not network design, but echo cancellers. Voice gateways often implement echo cancellation DSPs on the same interface card that does the voice coding and packetization. When we use echo cancellers we also need to be able to bypass them. One echo canceller is a good thing, but you can get too much of a good thing. If you've already got an echo canceller upstream, then you don't want to have another echo canceller. One echo canceller cancels the echo. Two echo cancellers degrades the voice quality. An end-to-end connection that traverses more than one packet network (e.g. the networks of several different carriers) needs an echo canceller at the ingress point to the first packet network but not at the ingress points to subsequent downstream packet networks. For fax and for modem data, we don't need echo cancellers, so we need to be able to bypass them.

### 2.3.3 Summary

We have therefore decomposed the delay issue into two: control of echo and control of delay to below 150 ms. Echo is important to humans but not to machines. Delay is important for interactive applications with human users at either end, such as a regular phone call. However, it is unimportant for machines, such as fax and modems, and also for situations with a human at one end and a machine at the other, such as voice messaging, and streaming applications. Also during the course of a phone call digits may need to be

**Table 2.1** Controlling delay in a packet voice network

|  | Echo | Delay |
| --- | --- | --- |
| Control mechanism | Echo canceller | Network priority mechanisms and edge equipment design |
| Important for | Human users of networks with four/ two-wire interfaces or acoustic feedback | Interactive applications with humans at both ends |
| Unimportant for | Fax, modem, dialled digits | Fax, modem, dialled digits, voice messaging, streaming |

dialled, e.g. to interact with IVR equipment, and again delay is not important here. Table 2.1 summarizes our discussion of delay.

## 2.4  Delay Variability and Jitter

### 2.4.1  Sources of Delay Variability

Now that we've discussed delay, we move onto delay variability or jitter. Figure 2.16 shows a source sending out voice packets in a regular fashion. Each packet hits the switch buffers at a different fill level, so that they all get delayed by different amounts by the time they reach the destination. One packet experiences a certain delay. The next packet on the same connection going through the same switches encounters different traffic conditions in the network, and therefore experiences a different delay. Each packet encounters a different delay as it traverses the network giving rise to delay variability. As an example of the extent of delay variability, on the Internet between Illinois and California during weekdays, the average delay was 80 ms, with a standard deviation of 10 ms. The standard deviation measures the delay variability.

Another source of delay variability is the use of non-real time operating systems. Figure 2.17 shows voice packets arriving from the Internet to a machine with an operating system that is scheduling processing power among a word processor, a spreadsheet, a Web browser and an audio tool. Each voice packet must, in general, wait a different amount of time for the operating system to schedule the audio tool.

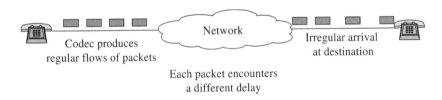

**Figure 2.16**   Jitter due to network

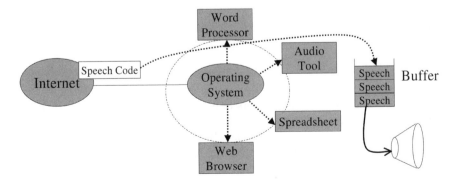

**Figure 2.17**   Jitter due to operating systems without real time support

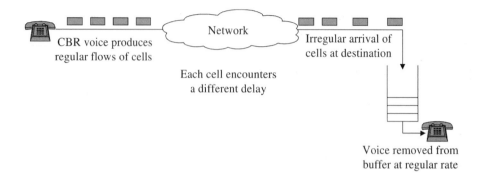

**Figure 2.18**  Compensating for jitter

### 2.4.2  Controlling Delay Variability

Delay variability can result in jitter in the reconstructed speech, which can be controlled by the use a buffer at the destination (see Figures 2.17 and 2.18).

The voice packets arrive at the buffer at irregular time intervals and this irregularity is absorbed by the buffer so that we remove the packets from the buffer at a regular rate. With a large enough buffer, it is really easy to kill off the delay variability completely. The larger the buffer, the more chance there is that the cells would sit in the buffer for long enough to absorb the delay variability. But one thing we don't want with voice is delay. So we've got a trade-off on the buffer size. The larger the buffer the better it is at absorbing the delay variability, but the worse it is at introducing delay in the end-to-end communication. So we've got to just hit it right on the size of these buffers so that they do absorb the delay variability but they don't introduce too much delay.

There is therefore a limit to how much delay variability can be absorbed by a jitter buffer before it starts to adversely impact total delay. This implies that it may be necessary to control delay variability introduced by the network to a level that can be handled by a jitter buffer. In ATM, for instance, the user is able to request a service with a limit on the cell delay variation. Jitter buffers can be tuned manually when the equipment is config- ured, or alternatively they can be set to automatically adapt to traffic conditions. Typically a jitter buffer would be set to 20–40 ms.

Delay variability introduced by the network needs to be controlled for human speech, including interactive telephone calls, voice messaging and voice response unit commu- nications. However, it is less of a concern for fax, dialled digits and modem data.

### 2.4.3  Summary

It is important to control delay variability in any application involving human speech. Delay variability introduced by the network or by non-real time operating systems in end- user equipment can be controlled by destination jitter buffers that absorb delay variability at the expense of introducing additional delay. In addition, ATM networks allow the user to request a specific value for the cell delay variation introduced by the network.

## 2.5   Voice Codecs

A codec is an algorithm, usually implemented in a DSP on a voice gateway or in software in a desktop computer, which converts between analogue voice and a digital representation of that voice. Codec stands for CODer/DECoder. Coding is done at the entry point to a packet network and decoding is done upon egress from the packet network.

There are two basic types of voice codec:

- waveform codecs;
- vocoders.

In this section we describe the major alternative codecs available and how they affect voice quality.

Clearly whatever voice codec is selected must be the same at each end of the communication. Gateways that transport voice over a packet backbone network negotiate with each other by standardized signalling messages as to which codec they are selecting (see Section 3.3.3).

### 2.5.1   Waveform Codecs

Figure 2.19 shows the analogue waveform of some voice being sampled every 125 μs, i.e. 8 KHz sampling (8000 times/s or once every 125 μs). The amplitude of the waveform is digitized using a certain 'quantization'. The more bits are used the more accurately can the amplitude be represented. This is known as Pulse Code Modulation (PCM) because we are generating a code at fixed time intervals or pulses. For instance, G.711 is the standard method of coding voice in a circuit switched network. It uses 8 bits every 125 μs, thus producing 64 Kbps of voice code. Audio quality can be improved by sampling more frequently and by using more bits per sample, as is done in coding music for CDs. In the telephony world, however, network clocks tick every 125 μs, so we retain the 8 KHz sampling rate.

In order to compress voice, we therefore need to use less bits per sample. One way of doing this is to code the difference between one amplitude and the next instead of coding the entire amplitude. Since the difference between two consecutive amplitudes is typically less that the actual height of the waveform, we would expect to get by with less bits per sample. This is known as Differential Pulse Code Modulation (DPCM).

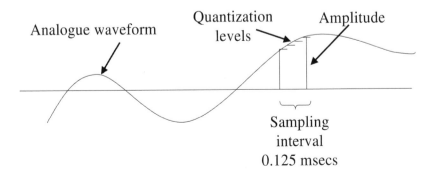

**Figure 2.19**   Waveform coding

If we are going to put enough intelligence into our codec to use DPCM, we might as well add some more intelligence to cater for the fact that some people speak louder than others and some networks attenuate the voice amplitude more than others. A quiet speaker is not benefiting from 8 bits of quantization if they never speak loud enough to use more than the first 4 bits. This implies the need for the codec to *adapt* to the *average* amplitude of the voice, and is called Adaptive Differential Pulse Code Modulation (ADPCM).

ITU-T standards G.722 and G.725 use ADPCM with 8 bits per sample to improve speech quality beyond what is normally referred to as toll quality. They produce 64 Kbps, the same as G.711, but since the code is a more efficient representation of the speech, the quality is improved. They are useful in speaker phone or video conferencing situations in which improved audio quality is required.

G.726 and G.727 use ADPCM to compress speech to 40, 32, 24, and 16 Kbps using respectively 5, 4, 3, or 2 bits per sample. In G.726, the sender and receiver must agree on which of these four coding methods to use and that method subsequently remains fixed for the duration of the call. In G.727, the code is divided into core bits, which are essential to decoding the voice, and enhancement bits which are nice to have to improve voice quality but we can get by without them. In the case of network congestion, the enhancement bits can be dropped by the network switches, resulting in variable bit rate operation, and correspondingly variable voice quality. A summary of waveform codec characteristics is given in Table 2.2.

When samples from a codec are transported over a packet network, they are grouped into packets. The number of samples per packet depends on the transport technology. For instance the specifications for voice over ATM Adaptation Layer Type 2 (AAL2) and voice over cable contain a specific range of possible packet sizes. The implementer chooses the packet size as a trade-off between delay and transmission efficiency. Waiting for more samples increases delay; however, it means that more information is transported per transport packet header, which improves transmission efficiency. The exact efficiency depends on the transport technology and some comparisons are given in Chapter 7; we summarize examples of possible impact on delay in Table 2.3. For example, the last row of Table 2.3 is obtained as follows: G.726 at 16 Kbps generates 2 bit samples (Table 2.2) 8000 times per second. If we choose to put 240 such samples in a transport packet, the size of the payload is $240 \times 2/8 = 60$ bytes. The delay is $240/8000 = 0.03$ s.

**Table 2.2**   Waveform codec sample sizes

| Codec | Kbps | Sample size (bits) |
|-------|------|--------------------|
| G.711 | 64 | 8 |
| G.722 | 64 | 8 |
| G.725 | 64 | 8 |
| G.726 | 16, 24, 32 or 40 | 2, 3, 4 or 5 |
| G.727 | 16, 24, 32 or 40 | 2, 3, 4 or 5 |

**Table 2.3**  Examples of impact of packetization on delay

| Codec (Kbps) | No. of samples | No. of bytes | Delay (ms) |
|---|---|---|---|
| G.711 | 80 | 80 | 10 |
| G.711 | 160 | 160 | 20 |
| G.726 (16) | 80 | 20 | 10 |
| G.726 (16) | 240 | 60 | 30 |

## 2.5.2  Vocoding

Unlike waveform coding, vocoding is based less on the analogue waveform than on the human vocal tract. It divides speech into voiced and unvoiced speech.

- Voiced speech resonates at a frequency that is characteristic of the human vocal tract, and is coded as the frequency plus the amplitude.
- Unvoiced speech is typically a consonant such as T or D and is represented as an amplitude without any specific frequency.

In this way speech is divided into phonemes which are the basic building blocks of speech. Thus, the word DOT contains three phonemes, an unvoiced D, a voiced O and an unvoiced T. Phonemes may last 10–100 ms, depending on the word being spoken, the intonation, the speaker and the language that is being spoken. Speech is therefore divided into 10–30 ms samples so that the voiced and unvoiced parts can be identified.

Vocoding achieves a significantly lower bit rate than waveform coding and requires more processing power in the codec. G.723.1, G.728 and G.729 are vocoding standards from the ITU-T which achieve 5.3, 9.6, and 8 Kbps, respectively, with a minimal degradation in quality compared to toll quality speech. There are three popular flavours of G.729: G.729A, G.729B and G.729E.

Voice quality from a codec is measured using Mean Opinion Scores (MOS), which use a scale of 1–5 on which 'toll quality' is 4. On a *perfect* network, i.e. a network with insignificant delay and loss, the MOS for alternative codecs is given in Table 2.4.

G.711 can be used as a reference codec, since it has been used for so long on circuit switched networks. It can be seen that vocoding introduces a slight degradation in voice quality compared to G.711, but is pretty close to toll quality (MOS = 4). Since G.711 uses 64 Kbps of bandwidth, vocoding has achieved a very significant bandwidth saving at minimal quality degradation.

On an *imperfect* network, however, where voice packets are delayed or lost, voice

**Table 2.4**  MOS for voice codecs

| Codec | MOS |
|---|---|
| G.711 | 4.3 |
| G.723.1 | 3.5 |
| G.728 | 4.0 |
| G.729 | 3.8 |

**Table 2.5** Vocoder packet sizes

| Standard | Kbps | Packet size (ms) | Packet size (bytes) |
| --- | --- | --- | --- |
| G.723.1 | 5.3 or 6.4 | 30 | 20 or 24 |
| G.728 | 9.6, 12.8 or 16 | 2.5 | 3, 4 or 5 |
| G.729A | 8 | 10 | 10 |
| G.729E | 12 | 10 | 15 |

quality degrades rapidly when vocoding is used. In particular, voice quality is highly sensitive to the number of *consecutive* packets lost since so much information is compressed into each packet.

Vocoders produce voice packets containing bits and bytes each of which represents some aspect of the voice code in a standard format. This is in contrast to waveform codecs which produce a stream of bits/bytes without any packetization. The packet sizes for the major vocoders are listed in Table 2.5.

As with waveform codecs there is a trade-off between transmission efficiency and delay. If the implementer decides to pack more than one voice packet into a transport packet, the percentage transmission overhead is reduced but the delay is increased. The extent of delay is given for some examples in Table 2.6. The delay is made up of the packet time from Table 2.5 and the look-ahead delay from Figure 2.12. The look-ahead delay only applies to the last voice packet that is grouped into a transport packet. For instance, the delay for four G.723.1 voice packets is $4 \times 30 + 7.5 = 127.5$ ms.

Vocoding gives us a trade-off between bandwidth and processing costs. It uses processing to save bandwidth. An 8 Kbps codec implemented in a low cost DSP in a voice gateway is therefore an economical alternative to paying for 64 Kbps of network bandwidth.

Since vocoders are designed to model the operation of the human vocal tract, they are unable to code fax, modem data and dialled digits. Fax and data modems use frequencies that are higher than those used in most human speech. Although a soprano may sing a higher note than a fax machine, we do not often listen to it over the phone. A call may start with G.729 coding of human speech and then the user starts to send a fax. The gateway needs to detect the fax squeal and switch to another codec, such as G.711, or demodulate the fax and convert it to data packets.

**Table 2.6** Examples of impact of packetization on delay

| Codec (Kbps) | No. of voice packets | No. of bytes | Delay (ms) |
| --- | --- | --- | --- |
| G.723.1 (6.4) | 1 | 24 | 37.5 |
| G.723.1 (6.4) | 4 | 96 | 127.5 |
| G.729A | 1 | 10 | 15 |
| G.729A | 6 | 60 | 65 |

### 2.5.3   Silence Removal

In addition to the coding methods described in Sections 2.5.1 and 2.5.2, codecs also employ silence removal algorithms, also known as speech activity detection algorithms. These enable further reduction in bandwidth by transmitting only minimal information when the speaker is silent, e.g. between phrases (or talk-spurts) (saving 10% of the bandwidth) and also when listening to other people speaking (saving 50% of the bandwidth). The 8 Kbps codec, G.729, requires on average only $(1 - 10\% - 50\%) \times 8 = 3.2$ Kbps, since it employs silence removal. Speech activity detection algorithms are not perfect and can 'clip' the speech at the start of a talk-spurt, so that an initial phoneme is lost. Speech clipping is an important item to test in selecting a new codec.

When silence is removed, it is important to play out background noise at the destination instead of total silence. Telephone users are accustomed to hearing background noise and interpret silence to mean that the connection has been cut. Therefore, during silence periods, codecs typically generate Silence Insertion Descriptors (SIDs), which tell the destination the amplitude of the background noise. SIDs are sent much less frequently than voice packets and consume very little bandwidth.

Some codecs such as G.723.1 and G.729 incorporate silence detection. Others can be combined with a separate silence detection algorithm.

## 2.6   Speech Recognition

A major reason for maintaining good voice quality is the growth of network-based services based on speech recognition. Speech recognition basically falls into three types.

- *Speaker-dependent.* The system is trained to recognize the voice of one particular speaker, and, once it is trained, it can recognize a very large vocabulary of words. This has limited use in the public network which is more geared to applications that many people can use.
- *Word spotting.* The system searches for specific words within the sentences spoken by the user. This is used for collect calls, but can have some problems as illustrated in the following dialogue:

  – System: Bob is calling from Denver; will you pay for the call? Please say 'Yes' or 'No'.
  – User: I spoke to Bob yesterday, I'm not paying to speak to him again!

  The user will be charged the collect charge for the call since they said 'Yes' within the word 'Yesterday'.
- *Speaker-independent.* The system can recognize a small vocabulary of words from a range of speakers with different accents, intonations, etc. This can be used for a range of telephony applications including:

  – accessing flight schedules by speaking the source and destination cities;
  – selecting a restaurant by speaking the type of cuisine and location;
  – obtaining traffic reports by speaking the highway number and location;
  – obtaining stock prices from a portfolio of possible company names.

The information being accessed can be located on a Web server, thus providing integrated access both from a Web browser over the Internet and also over a voice network using voice recognition and text-to-speech software. A major development facilitating Web-based voice response services is VoiceXML [7].

## 2.7   Monitoring Voice Quality

When the communications manager of a company negotiates a SLA with a network operator, measures of accuracy, delay and delay variability can be included. However, the end user may be more concerned with speech quality, for which there are two major measures: intelligibility and MOS. Intelligibility refers to whether a human recipient of the speech can understand it. This can be objectively measured by asking human subjects to state what they heard during a trial. The MOS is a more subjective measure in which human subjects are asked to give their own evaluation of overall speech quality on a scale from 0 to 5, where 4 represents toll quality speech. P.861 [5] is an ITU standard that aims to relate MOS to more objective measurements, called Perceptual Speech Quality Measurement (PSQM). PSQM can be estimated by software and is therefore less costly than the use of human subjects. Another such method is Perceptual Analysis Measurement System (PAMS), which measures not only the speech quality but also the amount of effort the user has to spend in understanding the speech. These are essentially objective methods of estimating subjective speech quality which have been found to correlate very well with human estimates.

When performing tests there is a trade-off between using pre-recorded voice and using live conversations (Figure 2.20). Pre-recorded samples allow standardized testing since the same sample is used each time. It can be replayed several times while parameters are adjusted to improve quality. Live conversations have the advantage of incorporating interactivity between two speakers which is not present in the case of recordings. However, live conversations are different each time they are used. For one-way voice messaging or streaming applications, pre-recorded voice can be used. For general purpose interactive and one-way applications, in practice, an average can be taken of the results

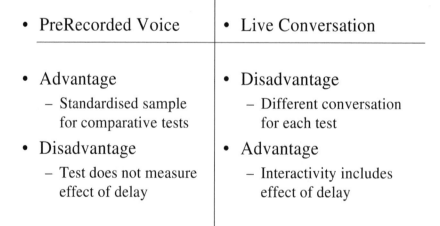

**Figure 2.20**   Assessing voice quality

with recordings and interactive conversations. The ITU-T standard P.861 can be used to test recorded speech instead of using human subjects.

MOSs are often quoted for specific voice codecs, assuming transport over a 'perfect' network. Network-based accuracy and latency measures of QoS are often quoted without relating them to speech quality. However, choice of a voice codec and choice of network-based QoS are related issues. In a network prone to packet loss, one would not use a codec that compresses a lot of speech into a single packet. In a network that delays the traffic, one would not use a codec that adds significantly to that delay for interactive voice traffic.

During network operation, test equipment can be used to monitor QoS in two ways.

- Active probes generate test voice traffic, e.g. a 20–30 s .wav file, and monitor the delay and loss introduced by the network between source and destination.
- Passive probes monitor the users traffic as it traverses the network, again measuring delay and loss characteristics. They can be set up to monitor individual calls or total network traffic.

Another type of monitoring is performed by standardized protocols.

- On an IP network, Real Time Protocol (RTP) is used to transport voice. For each RTP connection, there is also a Real Time Control Protocol (RTCP) connection, which monitors the performance of RTP.
- On an ATM network, standard OAM cells are interspersed with the user traffic to monitor the loss and delay characteristics of each virtual path and virtual circuit across the network.

In a connectionless network such as IP, it is not possible to monitor individual connections in the network since there are no connections, whereas in ATM and Frame Relay the connections can be identified by the connection identifiers in the packet headers. One way around this, in IP, is to monitor different *types* of traffic instead of monitoring individual connections. Traffic types can be identified by the use of standardized layer 4 port numbers. This is, however, becoming a less reliable method of identifying traffic types, as a result of the growth of the World Wide Web. Much Web-based traffic from mouse clicks to Web-initiated telephone calls is transported over Hyper Text Transfer Protocol (HTTP) for which the standard port number is 80. Thus, a very large amount of different traffic types is aggregated into port #80. QoS-sensitive traffic on the Internet can better be identified by the DiffServ Code Point (DSCP) instead of the layer 4 port number (see Section 4.4.2). This does not identify individual user connections but it does identify all traffic requiring a specific QoS, and is thus an appropriate way of evaluating whether that traffic is receiving its QoS.

Thus, the performance of a packet network can be monitored on an on-going basis during normal operation. The network manager can set traps in the network management system indicating the limit of acceptable performance. For instance, a trap could be set to alert the network manager if the end-to-end delay on a voice call exceeds 70 ms, or if the packet loss ratio exceeds 0.5%. Traps of these types send alarms to the network management system indicating a problem. Usually traps are set so as to give *early warning* of a potential problem. Thus, a trap of delay exceeding 70 ms could be set when the SLA requires the delay should not exceed 100 ms. The network manager is alerted to the problem in time to take action before it becomes noticeable to the customer. The kind of control actions possible are described in the next section.

## 2.8   Overall Control of Voice Quality

### *2.8.1   Tuning the Network*

Tests can be done on end-to-end voice quality and then certain parameters can be tweaked if it is necessary to improve quality.

Overall control of voice quality is a combination of network-based and edge mechanisms. The parameters that can be adjusted in order to improve voice quality are:

#### 2.8.1.1   Network-Based Mechanisms

- *The bandwidth of the access line.* A higher bandwidth line reduces the ingress/egress delay.
- *Priority level for delay.* Some network equipment, e.g. implementing ATM or DiffServ, allows a separate priority level to be set in order to reduce delay.
- *Priority level for accuracy.* Some network equipment, e.g. implementing ATM or DiffServ, allows a separate priority level to be set in order to improve accuracy.
- *Buffer management mechanisms.* Most queuing systems schedule the transmission of packets based on their priority level. Others take into account, not only the priority level, but also the length of time a packet has been waiting in line, thus reducing delay for packets that have already had a long wait.
- *Traffic engineering.* In an IP network, the regular Internet routing protocols select the path that is 'shortest' in some sense. This may result in a lot of traffic following the same path, thus resulting in congestion and delay. Traffic engineering allows the network manager to split the total traffic among alternate paths so as to alleviate congestion.

#### 2.8.1.2   Edge Mechanisms

- *Block size for FEC.* The longer the block size the more errors will be corrected, at the expense of an increased delay.
- *Block size for interleaving.* The longer the block size the more errors will be corrected, at the expense of an increased delay.
- *The number of sequence numbers to use for concealing lost packets.* A detailed discussion is given in Ref. [3].
- *The number of applications running on edge equipment with a non-real time operating system.* The less applications, the lower is the delay and delay variability.
- *Voice codec.* A higher bandwidth codec can in general improve voice quality at the cost of using extra bandwidth.
- *Silence detection.* Clipping of words can be improved by disabling silence detection at the cost of using extra bandwidth.
- *Number of voice packets/samples per transport packet.* Delay can be reduced at the expense of using more bandwidth, by reducing the number of voice packets or voice samples that are packed into a transport packet, e.g. IP, ATM or Frame Relay.
- *Length of jitter buffer at destination.* The jitter buffer length can be tuned so as to reduce delay while still absorbing jitter.
- *Echo cancellation time.* Echo cancellers operate by keeping a record of outgoing

speech for a certain period of time and searching for the same pattern in received speech. The length of time interval may need to be adjusted as users call different destinations for which the echo occurs after a different time.

### 2.8.2  Over-Engineering the Network

An alternative way of providing QoS is simply to over-engineer the network so that all traffic gets excellent QoS. Then it is unnecessary to deal preferentially with some traffic compared to other traffic. Over-engineering involves building a network with spare capacity so that there is 'no' congestion and traffic does not build up in switch buffers causing delay and/or loss. With developments in wavelength division multiplexing, bandwidth cost is low and arguably it costs less to provide additional bandwidth than it does to provide QoS via software and hardware mechanisms.

Over-engineering has been used successfully on ethernet LANs where bandwidth costs less than in the wide area network. However, there is a problem with over-engineering the wide area network: how much extra capacity do you provide? Some people suggest doubling the capacity. If we save 60% of the voice bandwidth by silence removal and then have to transport it on a network that is only 50% utilized, we haven't gained much! Also, it is necessary to apply the over-engineering not just to the voice traffic but to the total traffic on the network. Since traffic on a data-only network does not need QoS, it is currently over-engineered by less than 50%, let us say 25%. Suppose that during the busy hour, we have a T1 full of 64 Kbps voice and a T1 75% full of data. If we remove silence from the voice reducing the bandwidth by 60%, the combined traffic is 1.15 T1s. If we transport it over a network over-engineered to double the capacity, we now need 2.3 T1s, compared to 2 T1s when transporting the traffic separately. There is also the problem of 'focused overloads'. If a transmission link or a network node fails, the traffic is automatically re-routed, resulting is a significant increase in traffic on the alternate path, known as a focused overload. In order to continue to provide QoS, we need to over-engineer the network not just for regular traffic, but also for the traffic patterns that may result from re-routing around network failures.

Although over-engineering is an attractive 'quick fix' compared to the complexity of providing QoS, we are still left with a decision as to how much over-engineering is necessary and can easily end up with a greater bandwidth requirement on an integrated network than on separate voice and data networks.

## 2.9  Conclusion

Accuracy, latency, jitter and codec quality can be thought of as four musical instruments playing together in a quartet. As the composer and performers strive for beauty, depth and meaning in their creation, it would be vain to suggest that one instrument is more important than the others. Each has its role which must be played in concert with the others.

Accuracy is the piano. Latency and jitter go hand in hand as violin and viola. The flute can soar above the others as codec quality.

The piano should be played so that it does not detract from the violin and viola, and vice versa. They are contrasting instruments. The composer would have a tough time writing music with a major theme for the piano at the same time as there being another major theme for the violin and viola. Similarly, the network designer has a tough time building a network that is highly accurate and produces low delay and jitter. There is a trade-off between these design objectives.

The flute can soar above the other instruments so long as they do not interfere with it. The purity of its sound can be destroyed by percussive piano playing or a discordant note on the violin or viola. In the same way an excellent codec quality can be ruined by delay or information loss in the network.

Unity of sound can be obtained from our quartet by the composer (network designer) choosing balanced roles for the four instruments, and by the performers (network operators and users) tuning their instruments together and not over-emphasizing one instrument at the expense of the others. Can we achieve perfection by this kind of compromise? Perhaps not. But packet networks do not aim for perfect voice quality but instead for balanced voice quality that is acceptable to human ears accustomed to decades of telephone calls. It is balance that we seek to achieve in making our four instruments play together, not perfection. And that introduces another dimension: cost. A ticket to a concert with accomplished musicians costs more that a ticket to an amateur performance, and the price of admission to a packet voice network depends on the QoS provided.

## References

1. Floyd, S., Jacobson, V., Liu, C.-G., McCanne, S. and Zhang, L. (1997) A Reliable Multicast Framework for Light-Weight Sessions and Application Level Framing. *IEEE/ACM Transactions on Networking*, 784–803.
2. Hardman, V., Sasse, M.A. and Koouvelas, I. (1998) Successful Multiparty Audio Communication Over the Internet. *Communications of the Association of Computer Machinery*, **41**(3), 74–80.
3. Sriram, K., Lyons, T.G. and Wang, Y.-T. (1999) Anomalies Due to Delay and Loss in AAL2 Packet Voice Systems: Performance Models and Methods of Mitigation. *IEEE Journal on Selected Areas in Communications*, 4–17.
4. ITU-T Recommendation G.131 (1996) Control of Talker Echo.
5. ITU-T Recommendation P.861 (1998) Objective Quality Measurement of Telephone-band (300–3400 Hz) Speech Codecs.
6. ITU-T Recommendation G.168 (2000) Digital Network Echo Cancellers.
7. RFC 2198 IETF (1997) RTP Payload for Redundant Audio Data.

# 3

# Distributed Network Architecture

This chapter covers the distributed network architecture which is used to transport voice over a packet network. The main component in this architecture is the gateway, which we describe first. We then review the other components of the architecture in Section 3.2, and Section 3.3 is devoted to Media Gateway Control Protocol (MEGACO), which is used to communicate among the architectural components. Section 3.4 describes Session Description Protocol (SDP), which is used by MEGACO to give details of sessions being set up.

## 3.1  Gateways

Gateways are called different things by different people. The IP community calls them Media Gateways (MGs). The ATM community calls them Interworking Functions (IWFs). The Frame Relay community calls them IWFs or Frame Relay Access Devices (FRADs).

Gateways can be regarded as the functionality defined by standards bodies. This functionality can be implemented in existing equipment, e.g. a PSTN circuit switch, or it can be implemented in a dedicated piece of equipment which is itself called a gateway. Thus, the word 'gateway' and the equivalent terms given above may represent hardware/software functionality or it may represent the equipment in which that functionality is implemented. Equipment is generally sold by equipment vendors. Some packet switched network operators give away gateways to other carriers, in return for routing a given number of calls over the packet network operator's backbone.

### 3.1.1  Gateway Types

Gateways are available in implementations suited to three different deployments (Figures 3.1 and 3.2). The most powerful gateways are designed for the public network, sitting between the circuit and the packet parts of the network. Their distinguishing features are:

- reliability, following National Environmental Building Standards (NEBS) for power supply, air conditioning, and backup;
- throughput, generally requiring support for high traffic volumes;
- network management, generally requiring integration into the carrier's existing system

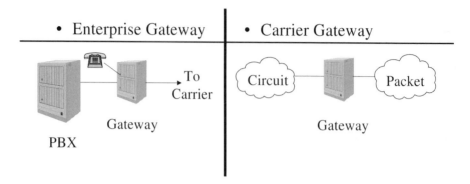

**Figure 3.1**   Enterprise and carrier MGs

using Simple Network Management Protocol (SNMP), and/or Common Management Information Protocol (CMIP).

The second most powerful gateways are suited to customer premises situations in businesses. They are sometimes called Integrated Access Devices (IADs) if they include data and/or video capabilities in addition to voice. Typically small/medium size companies buy enterprise gateways. *Very large* companies already get a significant discount on their phone bill from the public carrier because they have a large volume of traffic. The *small/medium* size companies who are not big enough to get those discounts can benefit from having an enterprise gateway at their site so as to reduce their telephony costs by using packet transport. These are companies which don't have a lot of data communication staff at every site and, therefore, the gateway needs to be remotely configurable. The characteristics required in this implementation include having a wide variety of interfaces, for instance:

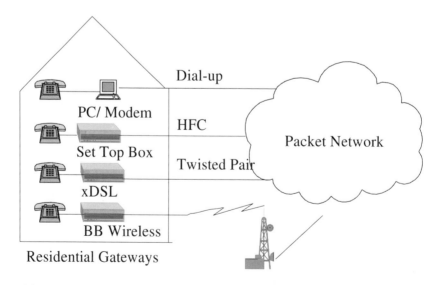

**Figure 3.2**   Residential MGs

- analogue and digital phones and PBXs on the customer side;
- Digital Subscriber Line (DSL), T1/E1, T3/E3 on the network side.

Third, there are residential gateways, which focus on stripped down functionality for low cost implementation. They may be used by individual residences, e.g. for one or two phone lines over a packet cable network, or by a small business requiring more phone lines. They are deployed in a variety of ways including:

- desk top computer;
- set top box;
- DSL modem;
- broadband wireless interface.

### 3.1.2   Gateway Functionality

Figure 3.3 illustrates the generic operation of a voice gateway. It takes in voice from, for instance, a T1, it looks at the individual channel in each time slot of the T1 and it analyzes the contents of that time slot. The first thing which it looks for is a modem squeal. This is a 2.1 KHz audio frequency and if there is no such squeal, then it looks for the presence of a Dual Tone Multi-Frequency (DTMF) signal, i.e. dialled digits during the course of a call. If there is no DTMF signal or modem squeal, then we assume that we have human speech on that time slot. We remove silence from the speech, and apply a voice compression algorithm and echo cancellation in order to generate voice packets which are sent off into the packet voice network.

On the other hand, if we do detect a modem squeal, there are two alternatives.

- The first alternative is to use a voice coding algorithm, possibly the same one which we are using for the human speech. But we cannot use a codec which compresses the

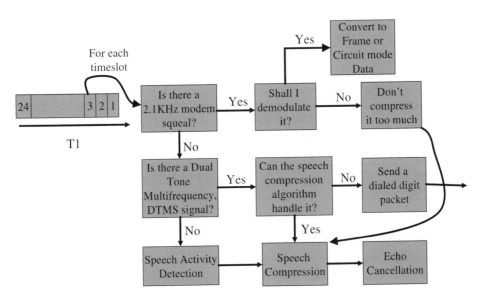

**Figure 3.3**   Processing voice in a gateway

speech too much, e.g. by vocoding. Modem squeals cannot be coded by codecs of that type (see Chapter 2). Typically we use a speech compression algorithm which generates 64, 32 or 16 Kbps.

- The second option for fax or data traffic over a modem is to convert it into frame mode data. In other words, we demodulate the fax or the data, convert it into data packets and then send it over the packet network as data packets.

The same thing happens with DTMF signals. They can be sent as dialled digit packets over the packet network instead of being coded using a speech compression algorithm.

We therefore need to be able to switch between one speech compression algorithm and another during the course of a call and also we need to be able to switch off the speech compression algorithm entirely in order to use dialled digit packets or frame mode data. We also must be able to switch off the echo cancellation in the case of a fax or a data modem call, since fax machines and data modems are not sensitive to echo.

This is the basic functionality of a MG. MGs are one of the fundamental building blocks of a packet voice network and we therefore now review additional features to look for in evaluating among alternative MG implementations:

- *The type of operating system and the reliability of the operating system.* This depends on whether the gateway is going to be implemented in a public network for which Unix is often used, or in the customer's premises, for which Windows NT may provide sufficient reliability.
- *The speech compression algorithms that are available.* G.711 is required for public network implementations since that is the voice coding on the circuit switched network. G.723 and G.729 are widely implemented, with G.726 and G.728 being optional in some gateways.
- *The fax detection options*, e.g. does the gateway allow us to demodulate fax and send it as a frame mode data or is there a speech codec which is suitable for sending fax? Can the gateway switch from speech to fax during a call or only at the start of the call?
- *The size of the jitter buffer at the destination*, which can range up to about 500 ms. It is important for a network operator to know whether the vendor of the equipment provides:

  - an unchangeable setting;
  - a default setting, which can subsequently be changed if required;
  - a tuneable setting, which the network operator has to set;
  - a setting which adapts dynamically to network conditions;
  - a service in which the vendor tunes the equipment to suit the carrier's network.

- *The type of echo cancellation*, particularly the length of time that is monitored to look for the presence of an echo, typically up to about 250 ms. The same options apply to setting the echo cancellation time as for the jitter buffer size above.
- *The signalling systems* available for setting up connections across the packet network. Typical signalling alternatives are H.323, Session Initiation Protocol (SIP), Simple Gateway Control Protocol (SGCP), Media Gateway Control Protocol (MGCP), MEGACO (another Media Gateway Control Protocol), RSVP, DiffServ, ATM signalling, for setting up Switched Virtual Circuits (SVCs) and AAL2 signalling for setting up AAL2 channels. The signalling software in the gateway needs to interwork with signalling software in other network nodes in the distributed architecture. It is not

sufficient to know that a vendor has implemented a certain standard protocol. It is also necessary to know whether interoperability tests have been performed with the vendors of other network nodes, since different vendors may have interpreted the standards differently and/or they may have added proprietary functionality to the standard.

- *The type of voice ports on the circuit switched side of the gateway.* Enterprise gateways and residential gateways need ports for plugging in analogue phones, PBXs and digital phones. In addition they need ethernet interfaces for connecting to the customers ethernets, for both data applications and ethernet phone and IP phone applications. For carrier implementations we need T1/E1, T3/E3 and STSn/STMn interfaces.
- *The interfaces on the packet side of the gateway.* Customer premises gateways need DSL, cable, and T1/E1 interfaces running ATM, IP and Frame Relay. Carrier gateways need T1/E1, T3/E3 and STSn/STMn interfaces running ATM, IP and Frame Relay.

Much of the functionality in a gateway is provided by DSPs including the voice coding, echo cancellation and packetization. Figure 3.4 indicates the number of Mega Instructions Per Second (MIPS) required at the time of writing to compress voice. As codec design and implementation improve this graph can be expected to come down, but we can always expect a trade-off between DSP MIPS and bandwidth. An example of this trade-off can be read off the graph for a 400 MIPS DSP. If we use this DSP to compress voice to 6 Kbps, we require 16 MIPS per voice channel, and therefore our DSP has a capability of $400/16 = 25$ voice channels. If, alternatively, we use the same DSP to code the voice at 8 Kbps, we require 11 MIPS per channel, so that the DSP has a capacity of $400/11 = 36$ voice channels. Thus, the capacity of a DSP in terms of the number of calls it can process depends on how much compression is required. A high throughput of calls can be achieved with little compression at the expense of more bandwidth per call in the packet network.

The trade-off between MIPS and bandwidth addresses the fundamental question of why we are compressing voice in the first place. Many people argue that developments in wavelength division multiplexing are reducing the cost of bandwidth so that it is unnecessary to save bandwidth by compressing voice. However, we are not compressing voice because bandwidth is expensive. We are compressing voice because bandwidth is more expensive that DSP processing. The cost of bandwidth is coming down and so is the cost of

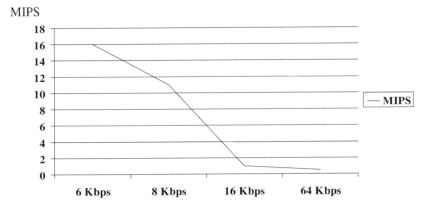

**Figure 3.4** DSP MIPS/Kbps trade-off

DSPs. It is economical to compress voice if the cost of that compression (i.e. the DSP cost) is less than the cost of the bandwidth saved.

### 3.1.3   Example Implementations

A gateway is not necessarily a newly designed piece of equipment. Different vendors have different approaches. It can be a new module on an existing piece of equipment, for instance:

- *A dial access concentrator.* This equipment has a pool of modems on it and as people dial in from their computers, typically for Internet access, they get one of these modems and the dial access concentrator concentrates the traffic into the Internet. A gateway can be implemented as voice cards on a dial access concentrator (Figure 3.5), allowing us to transport voice over the Internet.
- *A router.* Adding voice cards to a router also allows voice traffic to be sent over the Internet.
- *A circuit switch.* Circuit switches already have a lot of software to provide voice features and call set-up functionality. All that needs to be added is a packet voice interface.

These options allow vendors of existing equipment to modify it in order to provide voice gateway functionality. Alternatively, a new vendor can build a new piece of equipment as a specialized MG.

## 3.2   Distributed Network Architecture

### 3.2.1   Gateway Controllers

The most important components of the distributed architecture are the gateways and gateway controllers (Figure 3.6). The gateway controllers are sometimes called 'soft switches', even though they are not actually switches in the sense of switching the voice traffic from input ports to output ports. They are servers that control the gateways. The voice traffic never enters the gateway controller, instead it is switched by the gateway.

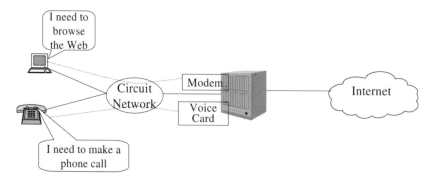

**Figure 3.5**   Dial-access concentrator implementation

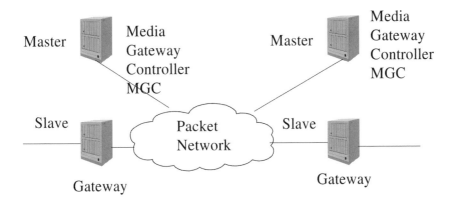

**Figure 3.6**  Distributed architecture

It is only the signalling messages that enter the gateway controller. Another term for a gateway controller is 'call control server', which appropriately describes its function.

Many signalling standards are available, including MEGACO (Section 3.3), which define the messaging between the Media Gateway Controller (MGC) and the MG. The packet network which is used to transport the voice traffic is also used to transport the signalling messages.

The gateway controller focuses on making routing decisions and communicating them to the gateway. The gateway focuses on coding the voice, providing special handling for fax, etc., and switching the voice from input ports to output ports. We now address the question of Why? Why not put all this functionality in a single component, as has been done with circuit switches for decades? Why distribute the functionality among different components of the architecture?

A common misconception about a distributed architecture like that illustrated in Figure 3.6 is that the boxes represent pieces of equipment. The boxes represent 'functions', i.e. jobs to be done. Each function could be performed in a separate piece of equipment or some functions could be combined into a single piece of equipment. Thus, although the reference architecture is distributed the implementation could be centralized. For instance, a circuit switch vendor could put a voice packetizing interface as one of the output ports from the switch, resulting in a centralized implementation.

However, most implementations distribute the functionality among different pieces of equipment, with the following advantages:

- A single gateway controller can control multiple gateways. This gives the network operator considerable flexibility in planning the network. As service is rolled out in new cities, only a MG needs to be installed for each point of presence, instead of having to install a complete switch with routing capability in each location. A single gateway controller can control additional gateways as they are rolled out.
- Reliability can be achieved with multiple gateways; if one gateway goes down, we can use another gateway.
- Of course a single controller is also a single point of failure. To solve this problem the architecture allows one gateway to be controlled by several different controllers. This is obviously advantageous from the point of view of reliability; if one gateway controller

goes down, we can use another gateway controller. Also controllers have a redundant power supply, and processing and storage modules that can kick in if the primary equipment fails.

- The gateway controller could be software running on a workstation or it could be the same software running on a circuit switch. Vendors of circuit switches have already implemented millions of lines of code for setting up calls. The distributed architecture allows that code to be re-used: either by using the circuit switch as the gateway controller or by taking the software and its hardware platform out of the circuit switch and selling it as a stand-alone product.

- Small vendors can specialize in a single component of the architecture, thus allowing them to enter the market in competition with large vendors, who may be able to offer all components.

- A standardized architecture with standardized protocols for communicating among the different components allows a network operator to design a multi-vendor network, thus reducing their dependence on a single vendor.

- Server software can be implemented on the latest hardware platforms being produced by the computer industry without having to develop hardware specifically for telecommunications applications.

An example of the effect of the distribution of functionality between a gateway and a gateway controller is its effect on Centrex. Traditional Centrex is an alternative to a PBX in which a customer outsources the PBX function to the public carrier (Figure 3.7). The customer has a multiplexer instead of a PBX and all calls (including calls internal to the customer premises) are switched in the public network switch. More access lines are required from the customer to the carrier with Centrex than with a PBX since internal as well as external calls need to be transported to the carrier. In Centrex both switching and call control are provided by the carrier. With a PBX, switching and call control are provided both by the customer and by the carrier.

A distributed architecture such as MEGACO can be used for Centrex and for a PBX (Figure 3.8). A distributed PBX is essentially a gateway controller at the customer site for internal calls with another controller in the public network for external calls. Centrex corresponds to just one call controller in the public network for all calls. This architecture

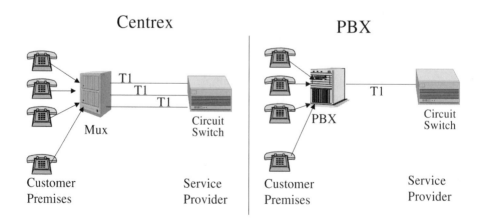

**Figure 3.7**   Centrex versus PBX

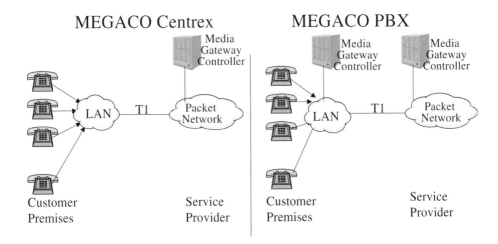

**Figure 3.8**   MEGACO Centrex versus MEGACO PBX

is different from traditional Centrex in that the customer LAN provides switching of internal calls in both the PBX and Centrex situations. Therefore, less access lines are required in the MEGACO Centrex case than in the traditional Centrex case. The effect of the distributed architecture can be summarized in Table 3.1.

**Table 3.1**   Centrex/PBX trade-offs in traditional and distributed architectures

|                       | Access bandwidth | Switching provided by: | Call control provided by: |
| --------------------- | ---------------- | ---------------------- | ------------------------- |
| Traditional Centrex   | High             | Carrier                | Carrier                   |
| Traditional PBX       | Low              | Customer and carrier   | Customer and carrier      |
| Distributed Centrex   | Low              | Customer and carrier   | Carrier                   |
| Distributed PBX       | Low              | Customer and carrier   | Customer and carrier      |

### 3.2.2   Signalling Gateways

The third type of component which we have in the distributed architecture is a signalling gateway (Figure 3.9), which interfaces to the SS7 network, brings in the signalling messages from the SS7 network and sends them to the MGC via the packet network. It may optionally perform some processing on those signalling messages; for instance, it may convert them from SS7 to MEGACO or some other format, or it may simply pass them on to the MGC unchanged.

As a call comes in from the circuit network, the voice traffic arrives on a time slot on the circuit network and the signalling associated with that voice call comes through the SS7 network. We therefore need the MGC to co-ordinate between the signalling coming through the signalling gateway and the voice coming through the interface to the MG.

The signalling gateway in Figure 3.9 is logically and physically separate form the MG since it interfaces to the SS7 network and the MG interfaces to the circuit switched

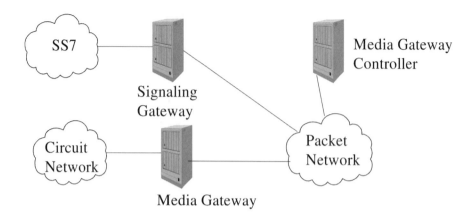

**Figure 3.9**   Signalling gateways

network. This is because signalling messages in the PSTN are carried on the SS7 network which is an overlay to the circuit switched network which transports the voice traffic. However, there are circumstances in which the signalling messages and the voice traffic both come over the same interface, for instance:

- an Integrated Services Digital Network (ISDN) Primary Rate Interface (PRI) line from a customer which carries 23 64 Kbps channels plus another 64 Kbps channel for Q.931 signalling;
- a DSL line from a customer which carries both signalling and packet voice;
- an SS7 F-link from the PSTN which is an exception to the rest of the SS7 network in that the SS7 signalling messages are transported on the same link as the voice traffic.

In these cases the interface terminates on the gateway, which therefore receives both the voice traffic (which it processes) and the signalling messages (which it passes on to the gateway controller).

We can distinguish between these two cases by referring to the signalling gateway in Figure 3.9 as an 'out of band' signalling gateway, since it handles signalling messages that are transported on separate links from the voice traffic. If the MG receives signalling messages of the above three types we refer to it as incorporating an 'in band' signalling gateway function, since the signalling messages and the voice traffic arrive on the same interface. Although an 'in band' signalling gateway could be regarded as a separate function from a MG it is unlikely that it would be implemented in a separate piece of equipment since both the signalling and the voice are arriving on the same interface. The 'in band' signalling gateway function is more likely to be software implemented in a MG.

### 3.2.3   Other Components

There are other components that can be optionally added to the distributed architecture.

- *Gatekeeper.* This is software which authenticates users and grants permission to access the network. A gatekeeper is defined as a separate function in H.323. It is usually implemented in the same piece of equipment as the gateway controller. A gatekeeper is generally required in a packet voice network to which end users have direct access. If the packet network is a backbone for transporting calls from PSTN access networks, the PSTN has already authenticated the users and there is no need for a gatekeeper in the packet network.

- *Feature server.* Voice features such as redial on busy are implemented in software on servers. They can be implemented on the same server as the gateway controller, or on a separate server. Separating the feature server function allows a vendor to focus on the development of special purpose feature software, e.g. for managing a corporate voice network. Another reason for implementing a separate feature server is to allow end-user customers or network operators to develop their own feature software accessing the feature server through an Applications Programming Interface (API). This allows network operators to compete with each other by developing their own software. It also allows them to become more independent of the equipment vendor, not having to order software development from the equipment vendor, but instead developing it themselves.

- *Audio server.* Servers optimized for playing audio messages usually need to be implemented separately from servers processing data traffic. Audio servers can be used to provide voice messaging and IVR capability.

- *Multipoint controller.* H.323 specifies a separate function for handling multi-party calls, known as a multipoint controller. MEGACO handles multi-party calls within the gateway and gateway controller and therefore does not require a separate component of the architecture for this purpose.

- *Billing server.* This may be incorporated on the same platform as the gateway controller and can offer a wide range of options including:

  - credit card;
  - calling card;
  - advertising revenue;
  - time of day;
  - usage-dependent;
  - bandwidth-dependent;
  - distance-dependent;
  - dependent on software application;
  - end-to-end delay;
  - information loss;
  - class of service;
  - chargeback to departments;
  - enterprise network usage;
  - off-network usage;
  - VPN encryption;
  - tiered pricing dependent on usage.

## 3.3   MEGACO

### 3.3.1   MEGACO's Position Among Other Signalling Protocols

Many signalling protocols are available for setting up calls across a packet network. H.323,

SGCP, MGCP and SIP were the first to be implemented for communication between gateways, gateway controllers and other components of the distributed architecture. This section focuses on MEGACO, since it is both an ITU [1] and IETF [2] standard. MEGACO is essentially the evolution of SGCP and MGCP and can be used on any packet network. SIP and H.323 are specific to IP networks and are therefore described in Chapter 4.

In addition to communicating between MGs and other elements of the distributed architecture, call establishment may require signalling between the nodes of the packet network in order to establish a path for the call. This is handled in IP networks by RSVP and DiffServ, which are described in Chapter 4. In an ATM network it requires ATM signalling for setting up SVCs and AAL2 signalling for setting up AAL2 channels (see Chapter 5).

MEGACO is the culmination of the developments in SGCP and MGCP, which it therefore replaces. It is simpler than H.323, which it therefore also replaces. It communicates between a gateway controller and a gateway. It works with SIP which operates between one gateway controller and another. It may call upon RSVP, DiffServ, ATM and AAL2 signalling as necessary within the packet network.

Figure 3.10 emphasizes the flexibility of MEGACO in allowing the implementer a choice of transport technology. The call can be set up either on an IP or ATM backbone depending on the interface on the MG. The MGC allows both transport technologies. Going over an ATM network we will typically use AAL2, adaptation layer number 2 (see Section 5.3.2). Going over an IP network, we typically use RTP/User Datagram Protocol (UDP)/IP (see Section 4.3).

### 3.3.2   Contexts

Each MG has a number of physical interfaces, each of which may carry a single voice call. Alternatively, a single physical interface may be divided into logical channels each of which can carry a call. For instance, an analogue twisted pair interface can be used for a single call. A T1 interface can carry 24 channels on different time slots. An interface into an IP network may incorporate UDP, in which case multiple calls are identified by UDP port numbers. We refer to these physical and logical interfaces on a MG as 'terminations'.

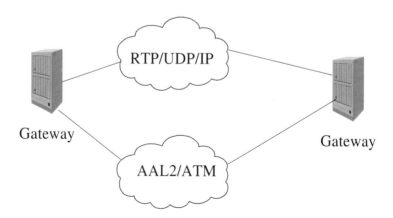

**Figure 3.10**   MEGACO use of alternative packet transport technologies

Each termination is capable of handling a single stream of voice traffic. A termination may be a modem, e.g. V.18, V.22, V.32, V.34, V.90 and V.91. A termination may be a multiplex, e.g. H.221, H.223, H.226 and V.75. Each termination can be send-only, receive-only, send/receive or loop-back.

MEGACO is based on 'contexts' in MGs. A context (Figure 3.11) is an association among terminations and specifies who hears whom in a voice call. MEGACO can also be used for video calls, for which a context specifies who sees whom. A context can have any number of terminations.

- An example of a context with only one termination is call waiting. The caller does not hear anyone.
- An example of a context with two terminations is a regular phone call. Each person hears the other.
- An example of a context with more than two terminations is a conference call. Each person hears all the others.

Call control can be effected by adding, subtracting and moving terminations among contexts. For instance Figure 3.12 illustrates a call between John and Janet, context C1. Sara also wants to speak with Janet and is initially put into a call waiting context C2. When Janet places John on hold to speak with Sara, the gateway controller instructs the gateway to move her termination from C1 to C2.

The context specifies the direction of traffic among the terminations. Figure 3.13 illustrates the case of a three-party call which starts with each party hearing and speaking to each other party, '1'. The three terminations in the context are labelled T1, T2 and T3. Then the gateway controller instructs the gateway to make T3 T1-unidirectional, '2'. The resulting traffic flows are indicated in '3'. Then the gateway controller instructs the gateway to isolate T3 and T2, '4'. The resulting traffic flows are indicated in '5'.

MEGACO commands relating to terminations in a context are:

- *Add/subtract/move* a termination to/from a context.
- *Modify* the description of a termination, e.g. the voice coding in use.
- *Notify* the gateway controller if an event occurs on a termination, such as a phone going off-hook, or digits being dialled.
- *Service change*. The gateway informs the controller if a new interface is added or if an interface goes out of service.

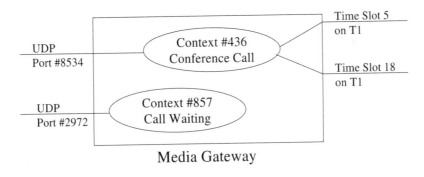

Media Gateway

**Figure 3.11**   A context in a gateway

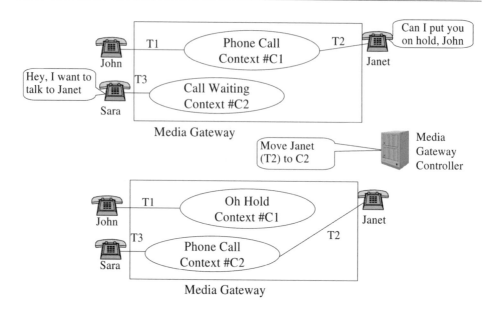

**Figure 3.12**   Adding/moving/subtracting terminations to/from contexts

## 3.3.3   Call Establishment

Setting up a call is the most basic function of MEGACO. The basic principles are illustrated in Figure 3.14, which indicates the sequence of operations '1', ..., '5'. We will refer to the media gateway controller as the MGC, the ingress media gateway as MG#1, and the egress media gateway as MG#2. First we review these basic principles and then we will go into more detail.

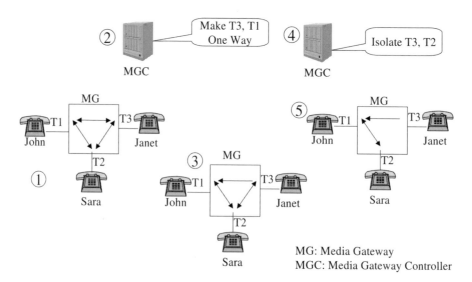

**Figure 3.13**   Communications allowed between terminations

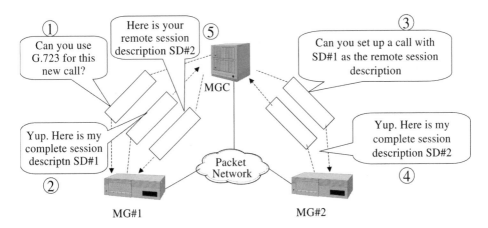

**Figure 3.14** Negotiating the local and remote descriptors

1. The MGC has received a call set-up message and has selected MG#1 as the appropriate ingress gateway. It requests G.723 coding on this call.
2. MG#1 responds that it can provide G.723 and returns a complete 'Session Description', SD#1, including the codec plus other details described below. The session description is formatted using SDP (Section 3.4).
3. The MGC passes this session description, SD#1, to the egress gateway, MG#2, for it to accept or modify as necessary.
4. MG#2 replies that it can accept SD#1, and adds some details of its own, in the format SD#2.
5. MGC passes SD#2 on to the ingress gateway MG#1.

In order to provide more detail on the call set-up scenario sketched above, we first review the operations for an IP network, and then add the functionality required for operation over ATM.

The operation over IP is illustrated by the 18 steps in Figures 3.15–3.20. The MEGACO command is illustrated by each numbered step in these diagrams. We take an example of a customer premises MG to which an analogue phone is directly attached.

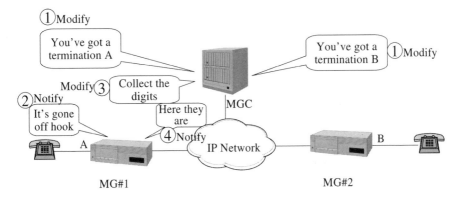

**Figure 3.15** Example of MEGACO IP call set-up: dialling

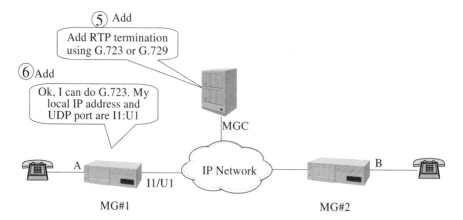

**Figure 3.16**  Setting up the context in the originating gateway

1. First the MGC configures the MG#1 by informing it to add termination A to a null context, i.e. a context with no other terminations. Similarly (not shown in the diagram) a null context with termination B is established in MG#2.
2. MG#1 notifies MGC that the phone on termination A has gone off-hook, i.e. someone has picked up the phone ready to make a call.
3. MG#1 does not have enough intelligence to know what to do next, so MGC instructs MG#1 to collect the digits dialled by the caller. This is an example of reducing the cost of the MG by implementing functionality in the MGC, and is particularly important for customer premises gateways since the market is price-sensitive and a large number of gateways are required, at least one per customer, whereas relatively few MGCs are required.
4. MG#1 notifies the MGC of the dialled digits.
5. The MGC determines that the call needs to be routed over the IP network and gives MG#1 a choice of two codecs to use. It also specifies the use of RTP, which is the normal way of transporting voice over an IP network (see Section 4.3).
6. MG#1 chooses a codec and informs the MGC of the UDP port number (U1) and IP

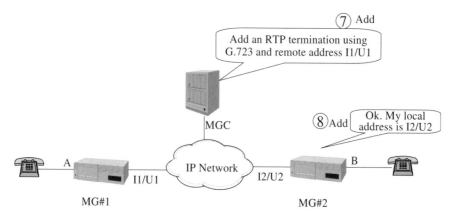

**Figure 3.17**  Setting up the context in the terminating gateway

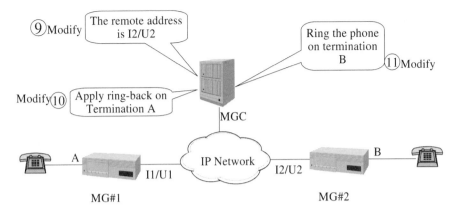

**Figure 3.18**   Finalizing the context in the gateways

address (I1) of the interface it has chosen to use to access the IP network. This information is conveyed using SDP. MG#1 adds U1/I1 to the context which previously just contained termination A.

7. MGC passes the session description on to MG#2, including the codec, and the remote termination U1/I1. It essentially says that it wants an RTP termination to be added using G.723 coding and it informs the destination gateway of the IP address and UDP port number of the ingress gateway.
8. MG#2 confirms that it can provide the necessary codec and returns an SDP message with its own UDP port number (U2) and IP address (I2). MG#2 adds U2/I2 to the context which previously just contained termination B.
9. MGC passes MG#2's information back to MG#1, so that now each MG knows the codec to use, plus the local and remote IP addresses and UDP port numbers.
10. MGC is now confident that the network is ready for the call. It tells MG#1 to apply ring-back on termination A, i.e. the caller hears a ringing tone.
11. MGC tells MG#2 to ring the phone on termination B, i.e. the called party hears the phone ring.

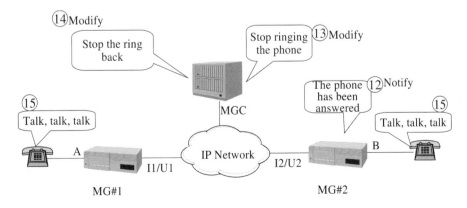

**Figure 3.19**   Answering the phone

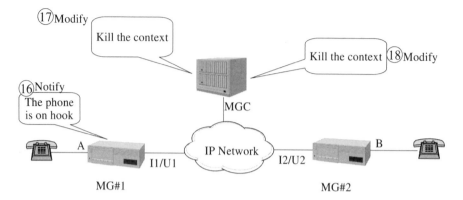

**Figure 3.20**  Tearing down the call

12. MG#2 notifies MGC that the phone on termination B has gone off-hook. Again we note that it does not have enough intelligence to know what to do about this.
13. MGC tells MG#2 to stop ringing the phone.
14. MGC tells MG#1 to stop playing the ring-back.
15. Finally the parties can talk over the network.
16. Tearing down the phone call happens when one of the gateways notices that the phone is on the hook. When the call is over, the phone on termination A is the first phone to be put down. MG#1 notifies MGC that the phone on termination A has gone on-hook.
17. MGC instructs MG#1 to modify the context that was used for the call. Termination A is returned to a null context. Termination U1/I1 is deleted so that the UDP port number U1 can be used for a subsequent call.
18. MGC instructs MG#2 to modify the context that was used for the call. Termination B is returned to a null context. Termination U2/I2 is deleted so that the UDP port number U2 can be used for a subsequent call.

That is the basic operation of MEGACO for call establishment when we are going over an IP network. When we're going over a connection oriented network like ATM, steps 5–9

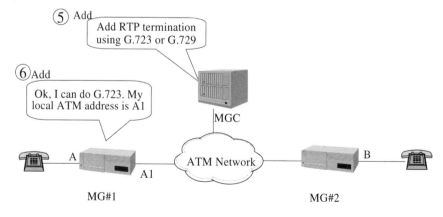

**Figure 3.21**  Setting up the context in the originating gateway

are changed, to take into account the connection oriented nature of ATM. A connection needs to be established across the ATM network, whereas in the above scenario we did not establish a connection across the IP network. If the IP network operator wants to be sure of good Quality of Service (QoS) on the IP network, a connection can be established over IP using RSVP or DiffServ, in which case the procedure is similar to what follows for ATM. The revised steps 5–9 are illustrated in Figures 3.21 and 3.22 and are as follows:

5. The gateway controller requests an ATM termination using G.723 or G.729.
6. The response from MG#1 confirms use of G.723 and informs the gateway controller of the ATM address (A1) of the output port which is being used to access the ATM network.
7. That information is passed on by MGC to the destination gateway MG#2. We note that MG#2 now knows the ATM address of MG#1 but not vice versa.
8. (A) MG#2 is therefore in a position to set up an ATM SVC, i.e. the connection is established from egress to ingress. MG#2 sends out ATM signalling messages backwards through the ATM network to the address A1. Private Network-to-Network Interface (PNNI) is a convenient ATM signalling protocol to use since it avoids bottlenecks in the network in order to ensure adequate QoS for the call. In other words, MG#2 is setting up an SVC through the ATM network to the ingress gateway. The egress gateway gives a connection number, S1, to this SVC and passes S1 on to the gateway controller.
8. (B) The ingress gateway, MG#1, confirms that the connection has been set up by responding with a connect message which is sent back to the egress gateway MG#2. Note that MG#1 has accepted an SVC *blindly*. It knows that this SVC will get it across the ATM network to an egress gateway, but it does not know which call context this SVC should be associated with.
9. The gateway controller passes the connection number S1 on to the ingress gateway, MG#1, and associates it with the appropriate call context. Now both gateways know the connection number of the SVC and which call to use it for.

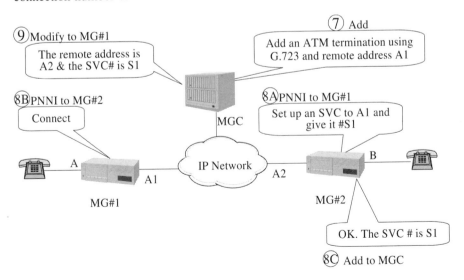

**Figure 3.22**    Setting up the context in the terminating gateway

The additional operations in the connection oriented network are that we have to set up an SVC and we have to assign a unique connection number to that SVC, which both gateways know about.

In summary, MEGACO is a reliable control protocol which runs between a MGC and a MG and it can operate over IP and ATM networks. It sets up contexts inside gateways and it also sets up sessions between one gateway and another, which are described using the SDP.

In discussing call establishment above we have referred to several protocols:

- MEGACO, which is the signalling protocol between the gateway controller and the gateway;
- SDP, which is used inside MEGACO to describe the bandwidth, the input and output port numbers, the codec, etc.;
- PNNI, which we use for setting up an SVC across an ATM network;
- RSVP and DiffServ, which are used to set up a connection across an IP network if we want to guarantee QoS on the connection.

Figure 3.23 shows how all these protocols fit together and also introduces the situation of two gateway controllers. SIP can be used between one MGC and another in order to locate where the destination user is. SIP is described in Section 4.5.2. We now describe SDP.

## 3.4   Session Description Protocol

SDP [3] is specified in RFC 2327 from the IETF. It can describe sessions which are unicast or multicast (Figure 3.24). It was originally designed to be used in conjunction with IP multicast, but can also be used in a unicast situation. Figure 3.24 illustrates how multicast can be provided from MEGACO and/or SDP. Unicast connections over the PSTN terminate in a multicast context in a MEGACO gateway. Another termination on this context interfaces to an IP network on which multicast is implemented in order to reach multiple destinations. Thus, IP multicast and MEGACO multicast combine to provide a very flexible multicast environment for the network operator. Multicast can be implemented using IP multicast, MEGACO multicast or both. SDP carries a description of the call in each case.

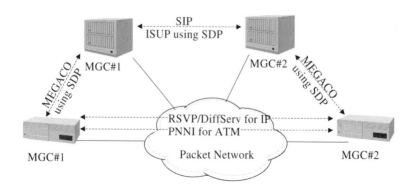

**Figure 3.23**   Fitting the protocols together

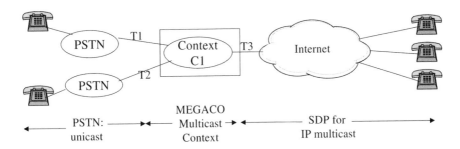

**Figure 3.24**   SDP RFC 2327

An SDP message contains various parameters to describe the call.

- The connection information, c, which includes:

  - the type of network, for instance IN stands for the Internet;
  - the address format to be used, e.g. IP version 6;
  - the address (unicast or multicast).

- The amount of bandwidth, b, in Kbps;
- The media information, m;

  - the media type: whether it is audio, video, application (e.g. whiteboard, i.e. application data presented to the user), data (e.g. executable data not presented to the user), or control (e.g. an additional conference control channel for the session);
  - the port number, e.g. UDP port number in the range 1024–655535 – even numbers are used for RTP sessions, with the next higher odd numbered UDP port for RTCP;
  - the transport protocol, e.g. UDP or RTP/Audio Video Protocol (AVP);
  - the media format, which, in the case of voice coding, uses a standard numbering system from Internet Assigned Numbers Authority (IANA) in which each of the voice codecs is allocated a corresponding number (Figure 3.25). A media format not included in the IANA list can be specified by using an AVP code in the range 96–127. IANA allows the user to specify dynamically the meaning of these codes in the 'additional information' parameter which we describe next.

- Additional information, a. Examples of additional information relevant to voice are:

| IANA # | Format |
|--------|--------|
| 0 | PCM |
| 2 | G.726 mu-law |
| 3 | GSM |
| 4 | G.723 |
| 9 | G.722 |
| 15 | G.728 |
| 18 | G.729 |

**Figure 3.25**   Sample media formats for voice (for a complete list, see www.iana.org)

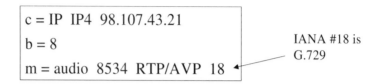

c = IP  IP4  98.107.43.21

b = 8                                                              IANA #18 is
                                                                   G.729
m = audio  8534  RTP/AVP  18

**Figure 3.26**   Example SDP specification

- the meaning of a user-assigned AVP code;
- the packet time, ptime in ms.

An example of an SDP specification is given in Figure 3.26. The connection information is the fact that we're using an IP network, the address is in IP version 4 format, and the destination address is in that format. The bandwidth is 8 Kbps. The session is audio, the UDP port number is 8534 using RTP/AVP and 18 is the IANA code for G.729. Optionally we could add a line to Figure 3.26 with additional information specifying the packet time of 10 ms:

$a =$ ptime :  10

An example of an SDP specification for the G.729E codec which does not have an IANA number is as follows:

$m =$ audio 3456 RTP/AVP 96

$a =$ rtpmap 96 G.729E/8000

The user-assigned AVP code 96 is defined in the additional information to mean G.729E with a sampling frequency of 8000 samples per second.

| CODEC | AVP code | Rtpmap | Ptime (msec) |
|---|---|---|---|
| | | SDP | |
| G.711 | 0 | | 10 |
| G.726* | 96-127 | G726-16/ 8000 | 10 |
| G.726* | 96-127 | G726-24/ 8000 | 10 |
| G.728 | 15 | | 10 |
| G.729a | 18 | | 10 |
| G.729E** | 96-127 | G729E/ 8000 | 10 |
| G.728E** | 96-127 | G729E/ 8000 | 20 |
| * Codec operating at different bit rates. | | | |
| ** Codec with different ptimes. | | | |

**Figure 3.27**   Example mappings of audio codecs to SDP. *Codec operating at different bit rates. **Codec with different ptimes

Other examples are given in Figure 3.27, and will be carried forward in Section 4.4.1 (Figure 4.17), to indicate the corresponding RSVP specification.

SDP specifies in these four parameters all the information which we need to set up a voice connection. The full SDP specification includes a number of other parameters as well, which are not essential for voice.

An advantage of SDP is that it uses ASCII text instead of binary, which makes it relatively easy to develop, test, debug and maintain.

## 3.5  Summary

We have now added many more instruments to the quartet of Chapter 2, with gateways, servers and gatekeepers all contributing to the packet voice orchestra. But were is the conductor? MEGACO provides communication between one instrument and another. It ensures, for instance, that ingress and egress MGs play in tune, but it does not provide any *central* control. The distributed packet voice architecture is more like a jazz 'big band' than a symphony orchestra. From time to time, individual instruments take the lead, showing off their own talents and making their own music while the rest of the orchestra plays only an accompaniment. The conductor determines which instrument improvises when, and may decide on the harmonies played by the rest of the orchestra, but within that framework the soloist does his/her own thing. When a call arrives in a packet voice network, the gatekeeper does its thing, then the call controller, then the gateways, each contributing its own specialized functions in harmony with the others. The main idea of a distributed architecture is that we do not need a conductor.

Part 1 of the book has introduced the packet voice 'big band'. It can play three main styles of music which are described in Part 2: Voice Over IP, ATM and Frame Relay.

## References

1. ITU Recommendation H.248 (2000) Gateway Control Protocol.
2. IETF RFC 2885, 2886, (2000) MEGACO Protocol.
3. IETF RFC 2327 (1998) Session Description Protocol.

# Part 2

# Packet Transport Technologies

Part 2 of this book describes how voice is transported on each of the three major packet transport technologies: IP, ATM and Frame Relay. This involves both engineering and business issues.

- *Engineering.* For each of the three transport technologies, we provide a brief introduction to the technology itself, with an emphasis on voice-related issues. We then take the voice code produced using the codecs described in Chapter 2 and indicate how it is mapped into the transport packets, using several alternative methods which are compared with each other. Quality of service provision on each technology is described with particular reference to voice transport. Connection establishment alternatives are identified including the use of permanent connections and signalled connections.
- *Business.* The market for voice over each transport technology is identified, and divided into:

  - desktop applications;
  - enterprise networking;
  - incumbent public carrier networking;
  - competitive public carrier networking.

Case examples are provided describing how each transport technology is applicable to each of these four markets. Provision of voice features is important in attracting customers to a voice transport service and we identify how additional features can be provided both from end-user and from network-based equipment. Cost and regulatory issues are also addressed.

Part 2 concludes with a chapter comparing the strengths and weaknesses of IP, ATM and Frame Relay for voice transport.

# 4

# Voice Over the Internet Protocol

## 4.1  The Market

This section identifies five segments of the market for voice over the IP and then describes the gateway features required.

### 4.1.1  Desktop IP Telephony

From the viewpoint of telephony, IP is the major protocol that can be used to deliver voice to the desktop. ATM deployment to the desktop is very limited, Frame Relay is an access and backbone technology, and when ethernet is used to the desktop at layer 2, it is generally carrying IP at layer 3.

A number of marketing studies (e.g. Refs. [1,2]) indicate the degree of user acceptance of telephony in association with a desktop computer. Although a significant number of people (20%) used personal computers to assist with telephony in 1999, that assistance was primarily with dialling calls from a regular telephone. Very few users (5%) were satisfied with the use of computer speakers and microphones for telephony. The majority prefer a regular telephone handset, which can be attached to a computer using a Universal Serial Bus (USB) port or attached to a LAN as an ethernet phone or IP phone.

In a survey of high-tech executives, analysts and IT managers, the major barriers identified to the use of voice over IP are:

- inferior voice quality;
- inferior service quality;
- diminished functionality.

This chapter shows how voice quality can be maintained, and service quality and functionality can be improved with voice over IP.

### 4.1.2  The IP PBX

The reluctance of some end users to use computer microphones and speakers for telephony has resulted in the flexible architecture known as the IP PBX, which gives end users a wide range of choices for edge equipment. Instead of the conventional PBX used at many business customer premises to switch internal and external circuit switched calls, an IP PBX provides the same functionality on a customer premises LAN, based on the distributed architecture discussed in Chapter 3. Figure 4.1 shows a customer premises IP network, typically based on an ethernet LAN, connecting a range of desktop and server equipment including:

### 4.1.2.1   Desktop Equipment

- IP phone. A digital phone incorporating voice coding and packetization functionality connected directly to the IP cloud, e.g. via an ethernet interface.
- Legacy analogue phone connected to the IP network via a voice gateway.
- Legacy analogue phone connected to the IP network via the user's desktop computer which incorporates gateway software.
- Legacy analogue phone connected to the IP network via the user's desktop computer which incorporates gateway software plus call centre software to allow calls and screen pops to be distributed to a pool of agents.
- Desktop computer with microphone and speaker or headset.

### 4.1.2.2   Servers

- Voice mail server connected to the IP cloud either directly or via a voice gateway.

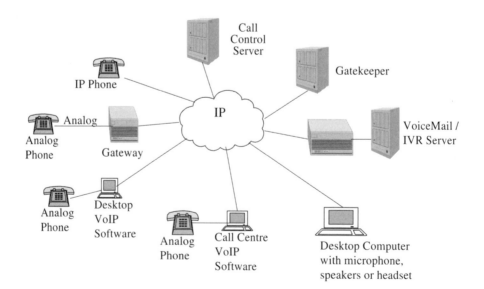

**Figure 4.1**   IP PBX distributed architecture

- Gatekeeper. An H.323, MGCP or MEGACO server providing authentication, authorization and access control functions.
- Call control server or softswitch, used to control the voice gateways using H.323, MGCP or MEGACO.

IP PBXs are a distributed architecture including some or all of the above components. The features they offer are similar to the features offered by the standard MEGACO architecture discussed in Chapter 3, including G.711 codecs for calls over the PSTN and G.723.1 and G.729 for internal calls. In addition there are several considerations that are specific to the PBX environment.

- *The maximum number of simultaneous calls.* This is dependent on the voice processing capabilities of the various gateways in the distributed PBX and also depends on the bandwidth available on the LAN.
- *The method of sharing LAN bandwidth with data traffic.* Some configurations use a LAN dedicated to voice without any issue of sharing with data. If the same LAN is used for voice and data, a priority mechanism is needed to ensure adequate performance for the voice.
- *The management functions.* Monitoring functions are required for utilization of link capacity, utilization of audio storage capacity, faults, and performance including latency, jitter and packet loss ratio.

An IP PBX can be distributed over more than one customer site. IP edge devices, such as IP phones and computers, at remote sites can be addressed from a call control server at a central site. This allows just a single call control server to be used for a corporate voice network interconnecting many sites. If the voice traffic is sent over the enterprise IP network which is also used for data traffic, considerable savings can be achieved from:

- requiring only one call control server plus backup instead of multiple conventional PBXs;
- traffic integration in the inter-site network.

IP Centrex is essentially a similar architecture to the IP PBX, where the call control server is located in the network instead of at the customer premises. IP Centrex is attractive to:

- customers wishing to outsource their voice networking requirements to a service provider;
- service providers wishing to offer a service based on a server without having to deploy any telecommunications equipment. Whereas the conventional Centrex is a subdivision of an Incumbent Local Exchange Carrier (ILEC) class 5 switch, an IP Centrex service can be based on a server attached to an IP network. All communications run over the IP network which can be provided by another network operator.

## 4.1.3  Voice over IP on Access

Moving on from the customer premises to the public network, many residential and business customers are using broadband access alternatives such as Asymmetric Digital Subscriber Line (ADSL) (Figure 4.2) for both data and packet voice. Chapter 9 describes how ADSL can be used with voice over ATM, which is suited to customers who do not use packet voice on their premises. However, with the IP PBX architecture, voice is already in IP format when it enters the ADSL access line, so that the IP packets need to be transported over ATM without repacketizing the voice into ATM cells. Alternatively, the ADSL terminating unit at the customer premises can connect to analogue phones as shown in Figure 4.2 and code and packetize the voice, acting as a voice gateway. Chapter 8 describes how voice over IP is standardized for the hybrid fibre coaxial cable environment of many cable companies.

In summary there are four packet voice alternatives for broadband access:

- voice over ATM over ADSL (see Chapter 9);
- voice over IP from an IP PBX over ATM over ADSL;
- voice over an IP gateway incorporated into the customer ADSL terminating equipment;
- voice over IP over hybrid fibre coax (see Chapter 8).

### 4.1.4   The IP Local Exchange

Figure 4.2 shows packet voice over an ADSL access line converted to G.711 64 Kbps in the local exchange for transport over the circuit switched network. In other words packet voice is hitting the local carrier from the customer side. It is also hitting them from the trunk side (Figure 4.3), since most of the initial implementation of voice over IP was in the long distance network (see Section 4.1.5).

Voice over IP on both sides of the local exchange is resulting in the emergence of IP local exchanges. Figure 4.4 shows on the left the conventional local exchange copied from Figure 4.2, and on the right the IP local exchange. The basic function of the IP local exchange is IP routing and switching, using conventional routers or the MEGACO distributed architecture. Optionally it can also include a voice gateway for access lines that carry non-packetized voice, e.g. G.711 coding in T1/E1 lines or analogue access lines.

### 4.1.5   Long Distance IP Telephony

In the short history of voice over IP, long distance has been the primary segment of the market and it has doubled or tripled each year.

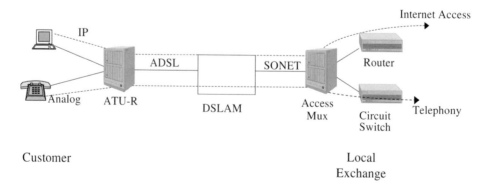

**Figure 4.2**   IP traffic on customer access lines. ATU-R, ADSL Transceiver Unit -- Remote

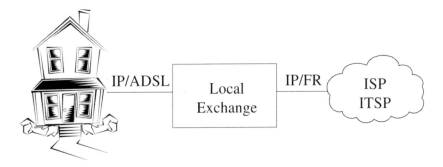

**Figure 4.3**   The IP local exchange

There is a broad range of carriers offering inter-city IP telephony. Many Internet Tele-phony Service Providers (ITSPs) offer voice over IP, both nationally and internationally. Internet Telephony Exchange Carriers (ITXCs) offer interconnection among different ITSPs, for instance in the case where a call originates on one ITSP and terminates on another. Some IP telephony providers focus on providing service to the retail market for individual switched calls typically from residential subscribers, some focus on the busi-ness market, offering PBX interconnect services, and a third group offer long distance and international services to local telephone companies. Some ITSPs offer service over the public Internet, but many regard the QoS available to be inadequate for a tariffed service, and have built their own managed IP networks. Most incumbent carriers fall into this category and transport voice over managed IP networks transparently, without the custo-mer being aware that a packet network is being used. Voice quality is a major element of customer satisfaction with incumbent carriers and they would not use a packet network if there was any perceptible degradation in quality to the customer.

An important issue for many ITSPs is whether to become a Competitive Local Exchange Carrier (CLEC). The advantage of doing so is that they would thereby gain access to the Signalling System No. 7 (SS7) network, which is used to carry the call establishment messages in the circuit switched PSTN. Thus, calls originating on the PSTN can be routed over the ITSP/CLEC network directly, and calls originating on the ITSP/CLEC network can be routed over another carrier that has implemented SS7. In

**Local Exchange with Circuit Switch**        **IP Local Exchange**

Internet Access

SONET   Router

Internet
Access

RT

IP

GW

Circuit   Telephony
Switch

Voice

**Figure 4.4**   Local exchanges compared

order to provide this interconnection, SS7 Point Codes are allocated to the ITSP/CLEC. A Point Code identifies their SS7 equipment uniquely for interconnection to the world-wide SS7 network. An ITSP that does not have an SS7 interface has to implement two-stage dialling, in which a customer first dials the number of the ITSP on their local telephone network or using toll-free dialling, and then receives a second dial tone for the call over the ITSP network.

The disadvantage to an ITSP of becoming a CLEC is that in the USA CLECs must pay access charges to the ILEC for calls originating or terminating on the ILEC network.

### 4.1.6 IP Gateway Features

Voice gateways are required in all of the five segments of the voice over IP market described above. They provide the interworking between the IP network and circuit switched or analogue interfaces. The functions of these gateways that are applicable to all packet technologies were discussed in Chapter 3. The functionality specific to IP that an implementer needs in a gateway and questions an implementer may wish to ask the gateway vendor are:

- Number of voice packets per IP packet

  - Does the vendor provide:

  - an unchangeable setting;
  - a default recommended setting;
  - a user tuneable setting;
  - a service in which the setting is adapted to the customer's needs?

- Voice call multiplexing in RTP

  - Is the method proprietary or standardized?
  - What is the maximum number of calls that can be multiplexed?
  - What is the overhead per call?

- Residential customer billing options

  - Does the billing system incorporate:

  - pre-paid and post-paid options;
  - cut-off of calls if payment runs out;
  - calling card creation and management;
  - individualized customer rate tables;
  - fax billing: per page or per minute;
  - gateway traffic analysis?

## 4.2 Cost Structure

Competitive pressures in most industrialized countries have resulted in significant reductions in long distance telephony pricing using the regular PSTN, even prior to the introduction of Internet telephony. Although IP telephony offers compressed voice and low cost switches, it is important to clarify whether these cost reductions are sufficient to result in price reductions to the end customer that are significant enough to migrate voice from a highly cost-efficient PSTN to an IP network.

One area in which IP telephony is a clear winner is in the international market where competitive pressures have been slower to reduce the PSTN pricing than in national markets.

National markets have seen a migration of traffic from the PSTN to IP networks due to the regulatory situation as much as due to the efficiencies of IP as a transport technology. In order to assess the relative importance of these two factors, we now present a costing analysis comparing PSTN and ITSP telephony in the USA.

Approximately 50% of the cost of a circuit switched, PSTN, long distance call is access charges. The remaining 50% is split between long distance costs and sales and administration. In the USA, the Federal Communications Commission (FCC) has classified IP telephony as an 'Enhanced Service', for which access charges are not collected by Local Exchange Carriers (LECs) from ITSPs. This results in a significant regulatory advantage to ITSPs over and above the advantages that IP packet transport provides. The 'enhanced service provider' exemption is not available if the ITSP becomes a CLEC.

The corresponding cost structure for ITSPs is:

- negligible access costs;
- gateway costs;
- long distance transmission and switching costs;
- sales and administration costs.

Voice compression reduces transport costs over the ITSP long distance network, but adds to the gateway cost, since additional processing is required the more the voice is compressed. The access charges are minimal for the ITSP, being simply the cost of a business line through which local calls are delivered to the ITSP office. Of course the utilization on this line is far heavier than for a typical business line, resulting in further effective subsidization of the ITSP by the incumbent local carrier. The sales and administration costs for ITSPs are typically lower than for incumbent telephone companies since they are younger, leaner companies with less management overhead.

The ITSP cost is therefore less than half of the PSTN cost, largely as a result of the enhanced service provider status of the ITSP. It is likely that this status is designed by the FCC to further stimulate long distance competition by giving the ITSPs a cost advantage that allows them to enter the market in competition with the much larger incumbent carriers. As such this status can be expected to be temporary, and to last until the FCC determines that ITSPs and IXCs can compete on a level playing field. Let us project that to occur by the year 2005, so that in that year both ITSPs and IXCs will pay the same access charges. During the period until 2005, all telecommunication costs (transmission, switching and voice coding) will also drop, let us assume by 50% compared to 2000. We also assume that the incumbent carriers become more efficient at sales and administration. Figure 4.5 shows the net effect of these changes to be a slight increase in ITSP costs in 2005 compared to 2000. However, the PSTN costs drop significantly to approximately the same level as the ITSP costs. The bottom line from Figure 4.5 is that the cost advantage that ITSPs currently enjoy may only be temporary.

As ITSPs prepare for the need to compete with IXCs on something other than cost, many regard the provision of additional services as being the important long term advantage that they have. Telephony services integrated with Web services are a natural combi-

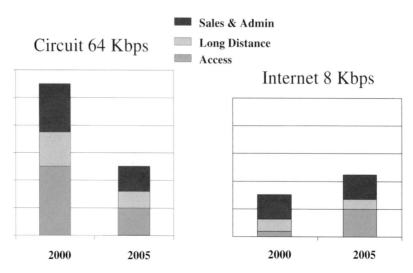

**Figure 4.5**   Evolution of circuit/Internet costs

nation for ITSPs and are more difficult for IXCs to provide over the circuit switched
network. Document sharing and video telephony are other examples of the way ITSPs
may be able to both attract and retain customers and also derive new revenue streams.
These services are described in Sections 4.6 and 4.7.

## 4.3   RTP/UDP/IP

Voice over IP involves more than just IP. Immediately above IP in the protocol stack is
UDP, and above that RTP. The functions performed by these protocol layers are summarized in Figure 4.6, with additional detail in Figures 4.7–4.9. The header structure is shown
with 4 bytes (32 bits) horizontally, following the convention which is useful for 32 bit
implementations. Not all detail is shown. We describe here only those features of RTP,
UDP and IP that are relevant to voice.

The RTP header is 12 bytes and provides:

- *Sequence number*. Since IP is connectionless, packets may become mis-sequenced due
  to traversing different routes in the IP network. The RTP sequence number can be used
  for resequencing. In addition lost packets can be detected at the destination from a gap
  in the sequence numbers, and action can be taken to compensate for the loss (see
  Section 2.2).
- *Timestamp*. When voice packets are not generated during silence intervals, the timestamp can be used to indicate the duration of the silence. The destination generates
  silence or background noise for this duration before playing out newly received voice
  packets.
- *Synchronizing Source ID*. In a multiparty conference the Synchronizing Source ID
  identifies which participant in the conference is currently speaking. This can be used
  for instance in a video conference to display the video of the person who is speaking.

RTP is supported by RTCP. RTCP packets are sent periodically to the same IP address

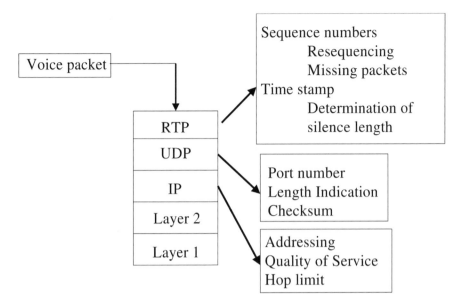

**Figure 4.6**  Transporting a voice packet

as the RTP session, but on a different UDP port number. RTCP provides additional information to participants in an RTP session:

- *QoS*. Errors, delay and jitter can be reported by the destination to the source using RTCP. In response, the source can change the encoding algorithm and frame packing.
- *Synchronization*. Audio and video streams are synchronized by information contained in RTCP packets sent from source to destination.
- *Identification*. Participants in a multiparty call can obtain each others e-mail address, phone number, etc. via RTCP.
- *Session control*. Participants in a multiparty call can leave the call by sending a BYE RTCP packet.

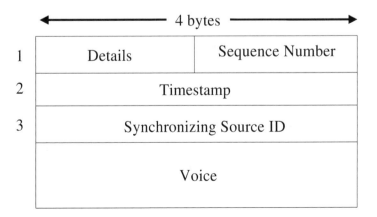

**Figure 4.7**  RTP

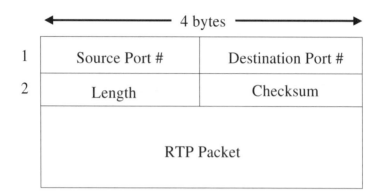

**Figure 4.8** UDP

The UDP header is 8 bytes and provides:

- *Source and destination port numbers.* UDP port numbers allow multiple voice calls to share a single IP source or destination address. This is essential for ITSP gateways, which simultaneously receive and send many different calls.
- *Length.* Voice codecs generate different length voice packets and hence different length UDP packets. The source sets the length indicator to inform the destination how long the packet is.
- *Checksum.* If errors are introduced during transport of the UDP packet they can be detected by recalculating the checksum at source and destination.

IP version 6 is being introduced to replace the previous IP version 4 for two reasons. First, the growth of the public Internet has resulted in a shortage of IPv4 addresses and IPv6 provides more address space. Second, some of the details in IPv4 are being elimi-

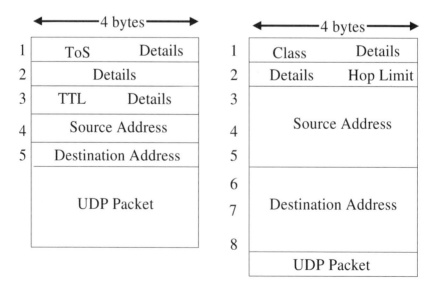

**Figure 4.9** IP

nated in IPv6 to provide a simpler protocol and hence facilitate hardware implementation. The IPv4 header is 20 bytes and the IPv6 header is 32 bytes. The services they provide for voice are:

- *Type of Service (ToS) IPv4; traffic class, IPv6.* One byte can be used to assign different QoS to voice traffic compared to other traffic as described in Section 4.4.
- *Time to live IPv4; hop limit, IPv6.* This number is decremented every time the packet traverses a network node. When it reaches zero the packet is discarded. This prevents misrouted packets from circulating for ever and also allows the sender of a packet to restrict how far the packet travels from the source. For instance, in a multicast, it can be used to restrict the recipients to a local area.
- *Source/destination address.* Addressing provides unique addresses for the source and destination of the packet across the Internet.

IP telephony therefore provides a number of features at the expense of a significant amount of overhead. The total overhead from RTP + UDP + IP is 40 bytes (IPv4) and 52 bytes (IPv6). For data applications payloads of up to 1500 bytes are typically used with TCP/IP for which the header is small by comparison. However, in the case of voice, the voice packet itself may be only 10 bytes, for instance for the G.729 codec, so that the header is much larger than the payload. Figure 4.10 indicates the very heavy overhead that a typical voice packet carries.

In practice, the overhead may be even greater due to the layer 2 protocol. For instance, the Point-to-Point Protocol (PPP) introduces an additional 8 bytes of overhead including the High-Level Data Link Control (HDLC) framing. ATM introduces at least 5 bytes. The layer 2 overhead has been omitted from Figure 4.10 because a range of different layer 2s could be used.

One application that is severely impacted by the heavy overhead is IP telephony over a modem. Let us take the case of a residential customer with a single POTS line and a 14.4 Kbps modem wishing to do IP telephony using G.729 at 8 Kbps. Although the codec speed is less than the modem speed, the modem cannot carry the resulting traffic. The 8 Kbps

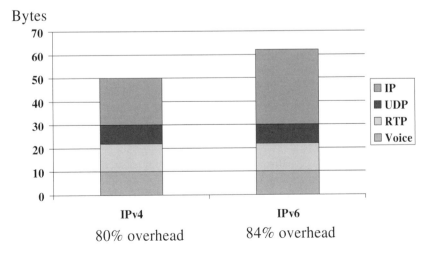

**Figure 4.10**  The voice packet

codec generates 10 bytes of voice every 10 ms. The additional 40 bytes of RTP/UDP/IPv4 header brings the total to 50 bytes every 10 ms, i.e. 40 Kbps. The minimum modem speed for transporting 8 Kbps of voice is 40 Kbps.

In summary, there are bandwidth savings and dis-savings in sending voice over IP. The savings come from the silence removal which reduces bandwidth requirements by 60–65%. However, Figure 4.10 indicates that the overhead increases the bandwidth requirements by between 5 and 6.2 times. The net result is a significantly increased bandwidth requirement.

There are three ways of overcoming the effect of the RTP/UDP/IP overhead:

- pack more than one voice packet from a single voice call into the same IP packet;
- pack more than one voice packet from different voice calls into the same IP packet;
- compress the RTP/UDP/IP header.

These alternatives are discussed in Sections 4.3.1, 4.3.2 and 4.3.3.

### 4.3.1   Frame Packing

If an IP packet is packed with more than one voice packet there are two advantages.

- First, the percentage of overhead is reduced. In the case of the 10 byte voice packets discussed above, an RTP/UDP/IP packet with $n$ voice packets has a proportion of overhead equal to $40/(40 + 10n)$ for IPv4 and $52/(52 + 10n)$ for IPv6.
- Second, the total number of packets traversing the Internet for a voice call is reduced in the ratio $1/n$. This reduces the load on the routers, since much router processing is done for each packet independent of the length of the packet, and router throughput is measured in packets per second.

In the case of multiple voice packets from the same voice call, an additional delay is introduced as a result of waiting for the packets to be generated by the codec. Following our example of G.729, with 10 ms packets and a look-ahead of 5 ms, the delay in assembling $n$ voice packets is $10n + 5$ ms. The delay at the destination is also increased, for instance because the jitter buffer needs to be expanded. Many commercial gateways allow packing of up to ten voice packets, introducing a delay which is unacceptable for interactive telephony. However, for streaming applications such as voice messaging, the additional delay from ten voice packets is acceptable and the bandwidth saving is significant.

For single voice calls there is therefore a clear trade-off between delay and bandwidth requirements.

### 4.3.2   RTP Multiplexing

In the case of multiple voice calls, one voice packet from each call can be packed into a single RTP/UDP/IP packet without increasing delay, since multiple codecs generate voice packets in parallel. However, a mechanism is required for identifying which voice packet belongs to which voice call. Figure 4.11 indicates that this implies yet another layer in the protocol stack: RTP multiplexing which sits above the RTP layer. Proposed methods [3,4] introduce 2 bytes of header per voice packet, but then allow the heavy overhead from RTP, UDP and IP to be shared among the multiple voice calls.

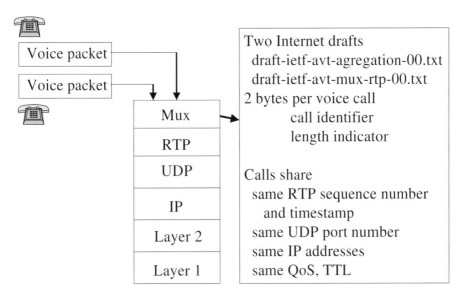

**Figure 4.11**   RTP multiplexing

All voice calls being multiplexed must share a common IP source and destination address, so that RTP multiplexing is suited to an ITSP environment in which many calls arrive and leave the IP network from PSTN access networks. All simultaneous calls going between a given pair of ITSP ingress/egress points can be multiplexed into a single RTP packet. Each call therefore shares a common RTP sequence number and timestamp, common UDP source and destination port numbers, and a common IP QoS. The bandwidth saving of multiplexing ten G.729 calls into a single RTP packet is shown in Figure 4.12. For IPv4 the overhead is reduced from 80 to 37.5%, and for IPv6 it is reduced from 84 to 42%.

An additional benefit of RTP multiplexing is to reduce the number of UDP ports required, since only one port is required for the multiplex instead of one port for each call. Without multiplexing, the vast number of individual voice calls could easily exceed the number of UDP ports on gateways. The maximum allowed by the UDP protocol is $2^{16}$ ports per IP address. RTP multiplexing therefore facilitates the scalability of IP telephony.

A typical RTP multiplexing implementation involves a number of calls arriving at a gateway from a circuit switched network, for instance in a T1. Let us suppose ten active calls need to go to a common IP egress point. The gateway applies voice coding to the ten calls, adds a multiplexing header to each voice packet, and packs them into an RTP payload until either a timer expires or the maximum size of the RTP payload is exceeded. At this point the RTP packet is passed down to the UDP layer in the protocol stack.

As RTP multiplexing is deployed in the network, the next stage is RTP switching in which network nodes switch voice packets according to the identifier in the multiplexing header.

### 4.3.3   RTP Header Compression

Once a voice call has been established across an IP network, an RTP/UDP/IP header is

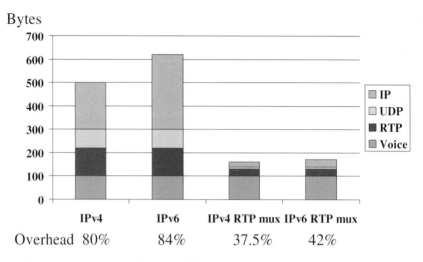

**Figure 4.12**   Ten voice calls with/without RTP multiplexing

generated for each voice packet in the stream of packets generated by the codec. Each of these packets has many properties in common, for instance they have the same IP addresses and UDP port numbers. The basis of header compression is to avoid repeating the same information on each packet. Some information may simply not be necessary, for instance the length of the packet can be specified by the layer 2 protocol and does not need to be repeated in the IP and UDP headers. The UDP checksum could be dispensed with as extra overhead without an important benefit. Some information such as the IP packet ID generally stays the same and we only need to send the differences from one packet to the

| Details | Length* | | |
|---|---|---|---|
| Packet Id *** | F | Offset | IP |
| TTL | Prot | Header Checksum* | |
| Source Address | | | |
| Destination Address | | | |
| Source Port # | Destination Port # | | UDP |
| Length* | Checksum** | | |
| Details | Sequence Number**** | | |
| Timestamp**** | | | RTP |
| Synchronizing Source ID | | | |

**Figure 4.13**   Are those headers really necessary? *Use layer 2; **optional; ***send differences only; ****send differences only if they change

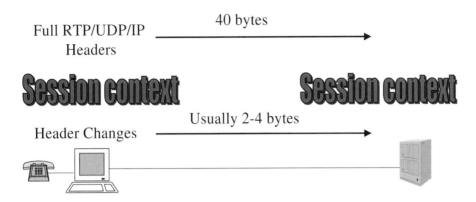

**Figure 4.14**   Unchanging info in the session context

next. Other information such as the RTP sequence number and timestamp do change from one packet to the next, but they change in a predictable way, e.g. the sequence number is incremented by 1. For this type of information we only need to send the differences if they change.

Figure 4.13 summarizes the ways in which most of the fields in an RTP/UDP/IPv4 header can be abbreviated.

Header compression [5] operates by sending the full RTP/UDP/IP header plus some other information at the start of a communication, which establishes a baseline 'session context' at either end of each link in the IP network (Figure 4.14). Thereafter, only information about changes to the context are sent. The session context consists of:

- full RTP/UDP/IP header;
- difference for IPv4 ID;
- difference for RTP timestamp;
- previous RTP sequence number.

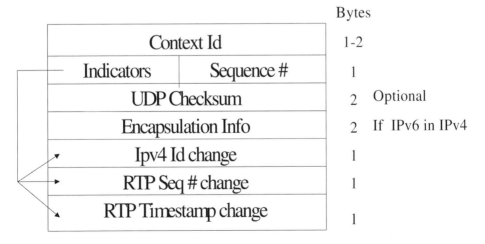

**Figure 4.15**   Format of change information

Since an IP link may be carrying multiple voice calls, routers need to maintain multiple session contexts. The appropriate session context for each call is identified by a context identifier which may be 8 bits (allowing 256 simultaneous calls) or 16 bits (allowing 65 536 simultaneous calls).

The format of the change information is shown in Figure 4.15. First the session context to which the change applies is identified, followed by a number of indicators showing which information has changed. The UDP checksum can be included if required, and encapsulation information is required if we are transporting IPv6 over IPv4. Each change packet has its own sequence number in order to detect lost packets.

The disadvantage of header compression is that over 40 bytes of storage is needed at the IP source and destination to store the session context. This represents the state of each IP telephony call, and needs to be maintained by the routers and updated as the change information arrives in the compressed header on each IP packet. Many people in the Internet community believe that the strength of a connectionless technology like IP is that it eliminates the need for maintaining the state of each customer's communication at routers within the network. State information limits the scalability of the network, and certainly in the case of telephony, a large number of calls can be anticipated, resulting in the need for a highly scalable network.

A possible deployment scenario that takes account of this problem uses the bandwidth saving capabilities of RTP header compression only in the access network where bandwidth is relatively costly. In the network backbone, high speed links render bandwidth a less limiting resource, so that header compression need not be used. This results in the need to maintain state information only in the access network, where routers are handling fewer calls than in the backbone. It is the backbone with its vast number of individual voice calls where scalability is an issue, and the use of RTP header compression only in the access network balances the need to save costly bandwidth while maintaining backbone scalability.

It is possible to combine header compression with RTP multiplexing. Header compression results in typically 2–4 bytes of overhead per RTP packet. Multiplexing adds 2 bytes per voice packet. Let us take the example of $n$ voice packets from $n$ different voice calls. Header compression on $n$ separate RTP packets results in a total between $2n$ and $4n$ bytes of overhead. Header compression plus RTP multiplexing results in a total of between $2 + 2n$ and $4 + 2n$ bytes of overhead. If multiplexing is used there is the additional processing overhead of ensuring that all calls have the same IP destination.

## 4.4  IP QoS

In IP there are two methods of providing QoS. The first uses RSVP and is otherwise known as Integrated Services. The second is known as DiffServ or Differentiated Services.

### 4.4.1  RSVP

RSVP involves a call set-up process in which the users request bandwidth and QoS. The routers in the network respond as to whether they have sufficient resources. Processing is required by each router en route from source to destination. If users and network agree, the call is established involving the maintenance of call state information in network routers. When the source starts to send packets to the network they are monitored in order to ensure that the source does not exceed the agreed upon traffic specification. Each packet has attached to it a flow identifier, which is analyzed by each router which it traverses so as to identify the packet as one for which bandwidth has been reserved.

There are two modes of RSVP:

- *Controlled load.* The user can request bandwidth but not QoS. The network will behave as if it was lightly loaded, but there is no quantitative guarantee of QoS.
- *Guaranteed.* The user can request bandwidth plus end-to-end delay. Delay is the only QoS parameter that can be specified in RSVP. It is not possible to specify any accuracy parameters; however, accuracy specifications are not of major importance in voice applications.

The source requests resources from the network. The network indicates to the destination what resources it has available. The destination may accept traffic with a lower resource requirement than originally requested by the source. RSVP is therefore suited to multicast applications since different destinations may accept different QoS. An interactive participant in a multicast may require a tight end-to-end delay, whereas a passive listener to the same multicast may be unconcerned about delay. Each can obtain the delay they need.

Bandwidth is specified in RSVP by a 'peak rate', a 'bucket rate' and a 'bucket depth', which are illustrated in Figure 4.16. Traffic policing at the ingress point to the network implements a 'leaky bucket' algorithm that allows traffic into the network at rates up to the 'peak rate' so long as the virtual bucket does not overflow. The maximum burst by which the traffic may exceed the 'bucket rate' is given by the 'bucket depth'. Thus, the long term rate at which traffic enters the network is the 'bucket rate' with short term bursts up to the 'peak rate'. ATM has a similar traffic policing mechanism.

The source and destination negotiate the call set-up parameters using PATH and RESV messages to exchange the TSpec and RSpec parameters listed in Figure 4.16. The source

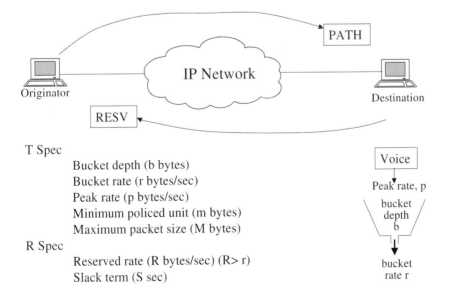

**Figure 4.16**   Standard RSVP RFC 2205

sends out a PATH message requesting a certain Reserved Rate, $R$, which is generally higher than the Bucket Rate, $r$. The reason for requesting $R > r$ is to reduce network delay. Each router and switch between source and destination examines the PATH message and estimates the delay that it would cause to traffic if the connection specified in the PATH message was established. These delays at each network node are accumulated as the PATH message traverses the network. Some delays, $D_i$ (in the $i$th network node), are fixed delays, independent of the bandwidth $R$. Others depend on the bandwidth according to the formula:

Nodal bandwidth-dependent delay $= C_i/R$

By the time the PATH message has reached the destination, it contains the cumulated total of all these nodal delays, $C_{tot}/R$ and $D_{tot}$, so that the destination can calculate the end-to-end delay as:

Total delay $= b/R + C_{tot}/R + D_{tot}$

If the destination requires a shorter delay, it is necessary to increase $R$, which the destination specifies in the RESV message. If the destination can accept a longer delay, it can inform the network about this by setting $S$ in the RESV message according to:

Acceptable delay $= S + b/R + C_{tot}/R + D_{tot}$

In return the destination would expect the network operator to give a discounted price. This allows different users to log into the same multicast with different delays and different prices.

An RSVP connection involves a sequence of network nodes (switches, routers) between source and destination. Once the RSVP connection is established, the stream of subsequent packets follows the path through these network nodes. This differs from the normal operation of the Internet, in which packets to the same destination do not necessarily follow the same path. RSVP has essentially imposed a 'connection' on the connectionless Internet.

The above procedure established an RSVP connection between a source and a destination, allowing one-way communication. Two-way communication requires a duplicate procedure involving a PATH message being sent by the destination and a RESV response from the source. In general the reverse path takes a different route across the Internet, and experiences a different delay.

RSVP parameters for voice traffic need to be related to the parameters for the voice codec, which are often conveyed at call set-up time using SDP (described in Chapter 3). Examples of constant bit rate parameter settings are given in Figure 4.17. For reference the codec parameters are given in Tables 2.2 and 2.5 of Chapter 2. An example of how to interpret Figure 4.17 is the rows for G.726. The SDP parameters indicate that this is a 16 Kbps codec with a packet time of 10 ms. To translate that information into RSVP, we first calculate the IP packet length, taking into account the RTP/UDP/IP overhead of 40 bytes. In these examples we are using the uncompressed header.

Packet length $= b = m = M = 16 \times 10/8 + 40 = 60$ bytes

We therefore have 60 bytes every 10 ms. The number of bytes per second is:

$r = p = (60/10) \times 1000 = 6000$ bytes/s

It is important to note that RSVP bandwidths are in bytes/s, whereas codec bandwidths are specified in bits/s.

Figure 4.17 gives two rows for G.726. The second row illustrates the effect of using G.726 at 24 Kbps instead of 16 Kbps. Figure 4.17 also gives two rows for G.729E, both at 12 Kbps, but with different packet times: 10 and 20 ms, respectively. A 10 ms packet time results in only 15 bytes of voice code being transported with a 40 byte RTP/UDP/IP header, resulting in an RSVP bandwidth requirement of 5.5 Kbytes/s. This can be significantly reduced to 3.5 Kbytes/s by packing two G.729E voice packets into one RTP packet resulting in a 20 ms packet time.

RSVP is similar to ATM QoS. Both allow the bandwidth and the end-to-end delay to be specified quantitatively. Both require connections to be established across the network and call state to be maintained in network equipment. Both require network equipment to participate in connection admission control, connection establishment and giving priority to packets for which guarantees have been given. However, ATM is optimized for point-to-point connections, whereas RSVP is optimized for multicast.

The connection oriented nature of ATM implies that network operators expect a significant amount of processing in network nodes to set up connections and deliver QoS. In the connectionless world of IP, less load is expected on network equipment, leading to the criticism of RSVP that it will not scale to situations such as telephony where a large number of separate connections need to be managed. These criticisms have lead to the development of DiffServ which maintains the connectionless nature of the Internet.

### 4.4.2 DiffServ

In DiffServ the QoS is specified in each individual IP packet that traverses the Internet. Six bits in the IP header, known as the DiffServ Code Point (DSCP), are used for this purpose as indicated in Figure 4.18. DSCP is a use for six of the bits in the IPv4 ToS field and the

| | SDP | | | RSVP | |
|---|---|---|---|---|---|
| CODEC | AVP code | Rtpmap | Ptime (msec) | b, m, M (bytes) | r, p (bytes/sec) |
| G.711 | 0 | | 10 | 120 | 12,000 |
| G.726* | 96-127 | G726-16/ 8000 | 10 | 60 | 6,000 |
| G.726* | 96-127 | G726-24/ 8000 | 10 | 70 | 7,000 |
| G.728 | 15 | | 10 | 60 | 6,000 |
| G.729a | 18 | | 10 | 50 | 5,000 |
| G.729E** | 96-127 | G729E/ 8000 | 10 | 55 | 5,500 |
| G.728E** | 96-127 | G729E/ 8000 | 20 | 70 | 3,500 |
| * Illustrates a codec operating at different bit rates. | | | | | |
| **Illustrates a codec with different ptimes. | | | | | |

**Figure 4.17** Example mappings of audio codecs to SDP and RSVP. *Illustrates a codec operating at different bit rates. **Illustrates a codec with different ptimes

IPv6 class field. DiffServ can be implemented in part of the Internet known as the 'Diff-Serv domain' where the routers understand the DSCP.

The DSCP is used to specify the 'per hop behaviour' that the packet receives from a network node. This behaviour can include accuracy, delay and bandwidth, without any quantitative specification of accuracy or delay. For instance, a user can request a lower delay on some of their packets compared to the delay on other packets, but receives no quantitative guarantee of the number of milliseconds that the packets will receive.

There are three DiffServ service classes:

- *Expedited, otherwise known as Virtual Leased Line*. This is the best class of service in terms of accuracy, delay and bandwidth. It allows the IP network to simulate the behaviour of a constant bit rate leased line, and can therefore be used to carry PBX interconnect traffic.
- *Assured forwarding*. There are 12 assured forwarding per hop behaviours, with three levels of accuracy (referred to as 'drop precedence') and four levels of delay (referred to as 'probability of timely forwarding'). This is similar to allowing the user to specify three priority levels for accuracy, plus four priority levels for delay. Typically voice would be assigned a priority for delay above that of most other traffic.
- *Class selector*. There are eight class selector codepoints. The higher the numerical value of the codepoint, the better the per hop behaviour from the point of view of delay. Accuracy cannot be specified. Voice is particularly suited to this service class and is typically given a high class selector codepoint.

Leaky buckets are used for traffic policing (as in RSVP and ATM) on ingress. The leaky bucket has levels corresponding to 'committed burst' and 'excess burst'. Drop precedence is set according to which of these levels a packet is at (see Figure 4.19). If the packet encounters a congested switch downstream in the network, Figure 4.19 shows that it is discarded according to how congested the switch is (the fill level of the switch buffer) and

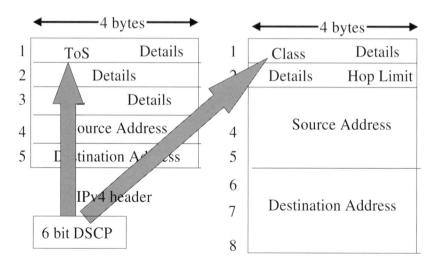

Figure 4.18   DiffServ

also taking into account the packet's drop precedence. If the switch is not very congested (switch buffer not very full), only the high drop precedence packets are discarded. At a more congested level the medium drop precedence packets are discarded as well. If the switch becomes very congested (switch buffer very full), all packets are discarded including low drop precedence packets.

The mechanism of setting the drop precedence according to the arrival rate of traffic requires the edge device to maintain information about the agreed arrival rates for all traffic streams from all customers entering the DiffServ domain at that ingress point. Each stream of traffic is policed separately with its own leaky bucket and burst size and arrival rate. However, once the traffic has been accepted into the DiffServ domain, the core routers do not need to maintain any information about individual traffic flows. They only need to read the DSCP on each packet and give that packet the appropriate per hop behaviour of accuracy (drop precedence) and delay (probability of timely forwarding). Thus, the call state information is maintained only in the edge routers, not in the core routers, whereas in RSVP, all routers need to maintain call state information about all RSVP flows. DiffServ can therefore be expected to scale better than RSVP.

In addition to traffic policing, edge routers perform other functions. When a customer registers for DiffServ service, they can set up a profile in the edge router which is used for their subsequent traffic. For example, the traffic profile can be specified according to:

- *Destination address.* Whenever the customer sends packets addressed to www.xyz.com they want assured forwarding with DSCP = 010100 and 56 Kbps.
- *DSCP.* Whenever the customer sends packets with DSCP = 101110 they want 1.5 Mbps expedited forwarding service.

Another function of the edge router is authorization to use the DiffServ service, i.e. checking whether the customer has subscribed to this service. Once the customer is authorized and wishes to send a new stream of traffic, a decision needs to be made as to whether to admit this new traffic into the network, i.e. connection admission control. A

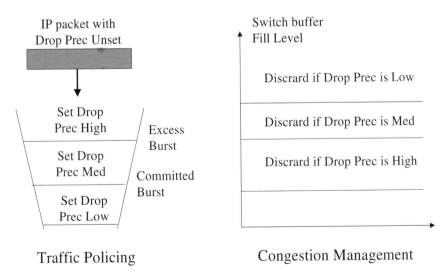

**Figure 4.19**   Drop precedence in DiffServ

**Table 4.1**  Location of DiffServ functionality

|                                      | Edge router | Core router |
|--------------------------------------|-------------|-------------|
| Authorization                        | Yes         | No          |
| Connection admission control         | Yes         | No          |
| Traffic policing                     | Yes         | No          |
| Packet classification and setting DSCP | Yes       | No          |
| Forwarding according to DSCP         | Yes         | Yes         |

summary of the functions performed by edge and core routers is given in Table 4.1, from which it can be seen that the majority of the functionality is at the edge.

### 4.4.3   Comparing IP QoS Alternatives

Table 4.2 compares RSVP and DiffServ from the viewpoints of the complexity of the implementation and the QoS provided. RSVP is more complex to implement, involving software upgrades to all network routers, and provides more sophisticated QoS in terms of quantitative guarantees on bandwidth and delay. DiffServ can be implemented by simply interpreting existing fields in the IPv4 and IPv6 headers in accordance with the DiffServ specification, thus facilitating hardware implementation. DiffServ allows the per hop behaviour of packets to be different relative to the behaviour of other packets, but does not provide any quantitative performance guarantees.

These differences indicate that DiffServ is more scalable than RSVP. A possible way of making the most of each technology is to implement RSVP at the edge of a network and DiffServ in the core. Thus, the scalability advantage of DiffServ is used in the part of the network with the most traffic flows. An end-to-end connection across the whole network including both the RSVP and DiffServ domains would not receive any quantitative guarantees of QoS. However, traffic just using the edge of the network would receive quanti-

**Table 4.2**  Comparison between RSVP and DiffServ

|                               | RSVP                                 | DiffServ                                           |
|-------------------------------|--------------------------------------|----------------------------------------------------|
| *Implementation complexity*   |                                      |                                                    |
| Connection admission control  | All routers                          | Ingress router                                     |
| Traffic policing              | Ingress router                       | Ingress router                                     |
| Connection establishment      | All routers                          | Not required                                       |
| Maintenance of call state     | All routers                          | Not required                                       |
|                               |                                      |                                                    |
| *QoS*                         |                                      |                                                    |
| Delay                         | Quantitative if using guaranteed service | Non-quantitative                               |
| Accuracy                      | Not available                        | Non-quantitative with expedited and assured service classes |
| Bandwidth guarantees          | Quantitative                         | Quantitative                                       |

tative guarantees from RSVP. An example of a network of this type combining the advantages of both DiffServ and RSVP is the hybrid fibre coax network discussed in Chapter 8.

## 4.5   IP Call Set-Up

IP is a connectionless technology. Communications across an IP network do not normally require permission to gain access to the network, nor do they require signalling to establish a connection. IP is designed so as to avoid the network nodes being involved in call establishment. For e-mail applications no connection is established across an IP network. For Web browsing applications a connection is set up between TCP software running in end stations, but not involving intermediate IP routers. Voice calls are different from data applications in four respects.

- They imply a relatively continuous stream of traffic.
- They have a requirement for QoS.
- In some cases the network operator may wish to bill the customer per call.
- The destination user may be mobile in the sense of having an IP address that changes dynamically. This may be because the user moves geographically to a new IP subnet, or it may be due to the fact that some ISPs allocate users new IP addresses each time they log on to the network. Dynamically varying IP addresses creates more issues for synchronous applications like telephony where both parties need to be present at the same time than it does for asynchronous applications like e-mail.

The first three of these requirements imply a need in some cases for users to request permission to make voice calls across the Internet. Permission may be required, both from the network as to whether there is capacity for the call, and also from the destination as to whether it can accept the call. The fourth requirement implies a need for network servers to provide 'find me' services.

When voice calls are set up between IP endpoints, signalling is involved from endpoint to endpoint or from endpoint to server; however, the intermediate routers in the network are not involved in setting up the call. The IP endpoint may be the end-user voice equipment or it may be a gateway into a non-IP network such as the PSTN. The reasons for signalling for voice call establishment are:

- The endpoints need to request and accept that the call be established.
- In cases where the endpoints must request permission to make a call, signalling is required to a server attached to the IP network in order to request that permission.
- Servers may be queried to perform address translation, e.g. between telephone numbers and IP addresses.
- Servers may be queried to find the current IP address of the destination user.

There are three major signalling protocols for setting up voice calls across an IP network: MEGACO, H.323 and SIP. MEGACO was described in Section 3.3 since it can be used with both IP and ATM networks. H.323 and SIP are described in the next two sections, since they are desinded for IP.

### 4.5.1 H.323

H.323 was originally developed by the International Telecommunication Union (ITU) [6] for real time multimedia communication over a non-QoS-capable LAN. It can also be used over the global Internet. It is based on two main standards, H.225 to set up a multimedia connection and H.245 to set up a single medium channel (e.g. voice) within a connection. There are two main servers in an H.323 network, the gatekeeper and the gateway as shown in Figure 4.20. The gatekeeper is mandatory and performs the following functions.

- It authenticates users as having permission to use the H.323 network.
- It translates addresses between IP addresses and E.164 (telephone numbers) and addresses of the form user@host.
- It determines whether the network has the capacity to accept a new call.
- It determines the routing of a call, e.g. whether the call should be directed to an intermediate gateway or whether it can go direct to the destination.
- It keeps track of the availability of bandwidth in the network.

The gateway is required if the H.323 network is connected to another network, e.g. the PSTN. It performs the functions described in Sections 3.1 and 4.1.6.

The other members of the H.323 family are shown in Figure 4.21, including:

- the terminal, which performs voice coding and packetization;
- the Domain Name Server (DNS), which enables a newly connected terminal to locate the gatekeeper;
- the Multipoint Control Unit (MCU), which acts as a bridge to provide multipoint conferencing.

We now provide a brief overview of H.323 call establishment for the case of a source and destination, both on the same H.323 network and both registered with the same gatekeeper. Other scenarios are possible, e.g. one of the endpoints may not be registered with a gatekeeper, or may be registered with a different gatekeeper. Multipoint communications are routed via the MCU. The reader is referred to the ITU standard [6] for the fully fledged complexity of H.323 call establishment.

Figures 4.22–4.24 illustrate the H.323 call establishment from Bob to Helen. The generic names Bob and Helen are intended to represent IP addresses, E.164 addresses or addresses of the form user@host.

Bob's first task (Figure 4.22) is to locate the gatekeeper for his H.323 domain, which he

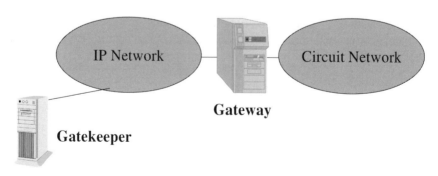

**Figure 4.20**   H.323 gateways and gatekeepers

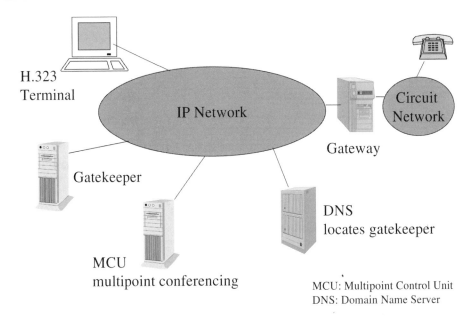

Figure 4.21   The H.323 family

does by querying the DNS. He registers with the gatekeeper and then uses H.225 to request permission from the gatekeeper to call Helen. The gatekeeper assesses the availability of capacity on the network and decides whether to grant permission. If Bob requires address translation from a non-IP format, the corresponding IP address is provided by the gate-keeper.

The connection establishment message to Helen may be sent to Helen by the gatekeeper on Bob's behalf (not shown on the diagrams) in which case it needs to maintain call state

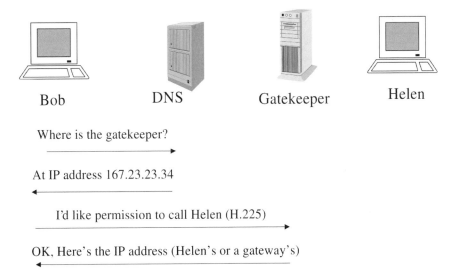

Figure 4.22   H.323 call establishment: address translation

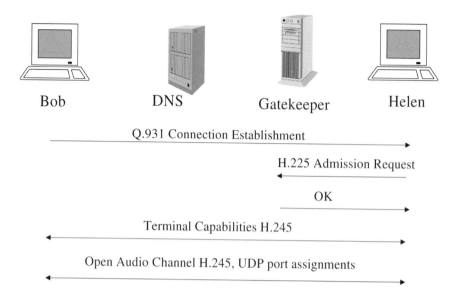

**Figure 4.23**   H.323 call establishment: setting up the connection

information. Alternatively it may be sent by Bob to Helen direct (Figure 4.23). If Helen
accepts the call, she requests permission to enter the network by sending an H.225 admission request to the gatekeeper. The gatekeeper authenticates Helen, and then Bob and
Helen can use H.245 to exchange information about their terminal capabilities, including
voice coding algorithms and frame packing options. Once the call has been established,
channels are opened within the call, for instance a voice channel. UDP port numbers are
assigned for each channel.

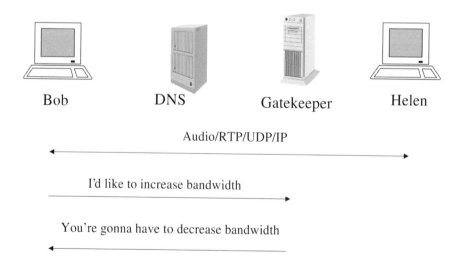

**Figure 4.24**   H.323 call establishment: modifying the connection

Call set-up is now complete and the parties can send and receive voice over RTP/UDP/ IP (Figure 4.24). During the course of the call, an endpoint may request to increase the bandwidth of the call. If the bandwidth of one channel is increased and another channel decreased in such a way that the total call bandwidth is not increased, no request is necessary. The gatekeeper can also request the communicating parties to reduce call bandwidth if network conditions require it.

H.323 call establishment is complex and the standard is continuously evolving, with version 2 in 1998 and version 3 in 1999. The use of an MCU for multicast is costly and does not exploit IP's lower cost multicast capabilities. For these reasons, the Internet community has developed a more streamlined protocol, SIP [7], which we discuss next.

### 4.5.2   SIP

SIP, like H.323 and MEGACO, is based on servers and aims at establishing a connection between participants in a call (Figure 4.25). However, SIP is simpler in only having servers whose main function is to locate the called party. Figure 4.26 shows one physical server which is a typical implementation incorporating three logical server functions, register, proxy and redirect, which will be described later.

SIP is designed to operate over IP. For calls in which some participants are on non-IP networks, MGs defined in other technologies, such as H.323 and MEGACO, are required. There is nothing to stop SIP being used on a non-IP network, such as ATM. In the same way, if a SIP ATM network needs to set up a call to a participant on another network, a non-SIP gateway must be used. Interworking between SIP, MEGACO and H.323 is there- fore an important requirement and equipment vendors often implement all three technol- ogies on a single hardware platform to facilitate interworking and to give a choice of technology on a call-by-call or customer-by-customer basis.

SIP is a terminal-to-terminal protocol. The server shown in Figure 4.25 is not required for all interactions. For instance, Figure 4.26 shows one terminal querying another term- inal as to the capabilities offered for voice communications. The phone does not ring and the server is not involved.

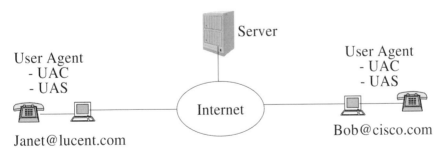

**Figure 4.25**   SIP, overview. UAC, User Agent Client – make a call; UAS, User Agent Server – answers a call

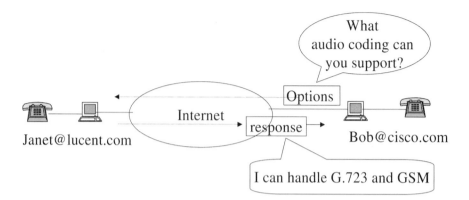

**Figure 4.26**   SIP: querying a callee's capabilities

At the start of a SIP session:

- a SIP terminal locates its local SIP server using regular DNS look-up;
- the server then authenticates the user;
- if the session is to be encrypted, the terminal and server exchange encryption keys which are to be used during the session. Symmetric key encryption, which does not have a heavy processing overhead, is generally used during the session. However, public/private key encryption, which is more secure, is used for authentication and for the exchange of the symmetric keys that are to be used during the session.

Locating SIP users is facilitated by the registration procedure illustrated in Figure 4.27. A user planning to be at a different address can be registered with the Registration Server together with a period of time during which the new address is valid. Address changes are for two reasons:

- the user may be moving geographically to a new location, on a different IP subnet;
- the user may be allocated an IP address dynamically each time they access the Internet. Dynamic address allocation is often used by ISPs with dial-up customers as a way of limiting the total number of IP addresses required. Different IP addresses are then required for those users currently on-line, not for every ISP customer.

There are two modes in which a caller can set up a call: Redirect Mode and Proxy Mode. Servers are defined in SIP for each of these two modes, although the software may well be implemented on a single hardware platform.

Figure 4.28 illustrates Redirect Mode. The caller issues an 'Invite' message to invite another user to participate in a call. The message is sent to the redirect server which checks its database for the current IP address of the person being invited. If it does not have a current address available, it can query another server, possibly initiating a chain of queries designed to locate the required IP address. Once the destination IP address is found it is passed back to the caller, who then issues another invite message directly to the invited party.

Proxy Mode is illustrated in Figure 4.29. The caller issues the 'Invite' message to the server in the same way as in Redirect Mode. The difference is that in this case, once the

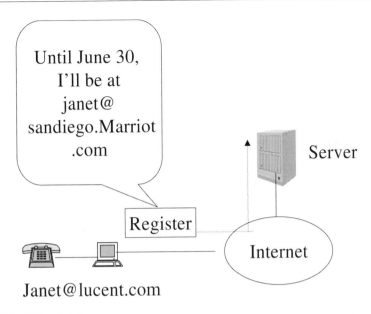

**Figure 4.27**  SIP portability

address of the invited party is determined, the Proxy Server passes the 'Invite' message on to that address.

In both Redirect and Proxy Modes, if the invited party wishes to accept the invitation, they reply direct to the caller, not to the server.

Advantages of SIP over H.323 and MEGACO are:

- The server does not maintain any ongoing information about the state of calls in progress. SIP is therefore a very simple protocol compared to H.323 and MEGACO in which call state information is maintained.

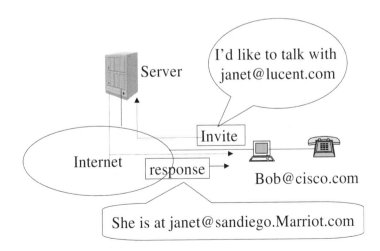

**Figure 4.28**  SIP redirect

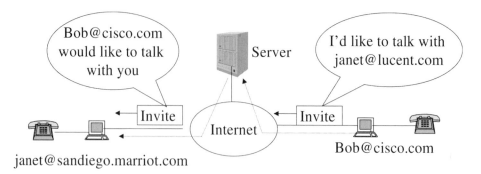

**Figure 4.29**   SIP: proxy mode

- SIP allows multiparty calls to be set up using IP multicast capabilities which therefore do not require the use of a multipoint controller as in H.323. The same 'Invite' message is used for invitations to multiparty calls as for point-to-point calls.
- SIP messages are text-based, as opposed to binary, resulting in ease of writing and debugging associated software. Another example of a text-based messaging protocol is HTTP, which has had extremely rapid market acceptance partly as a result of its simplicity. SIP is also being rapidly accepted by the industry.
- The intelligence is pushed to the network edge where processing capability is available in desktop computers. Only the location of invited parties, not the actual call set-up, is handled by network-based servers. The advantage of this is simplicity. However, network operators may be reluctant to support SIP since they loose control of the call set-up process. They therefore do not know whether calls are in progress and cannot bill for them.

## 4.6   Web Integration

In this chapter we have seen some of the technical challenges for voice over IP and the solutions available:

- excessive header overhead which can be solved by RTP multiplexing and header compression;
- QoS requirements which can be provided by RSVP and DiffServ;
- connecting to possibly mobile users which can be accomplished using H.323 and SIP.

IP telephony was implemented and sold as a commercial service before the above solutions were available on many IP networks. This resulted in a service that was profitable although it was not as efficient as it could be. However, we also saw in Section 4.2 that the profitability of IP telephony may be short-lived, due to the probably temporary nature of the FCC enhanced service exemption. It can be said that the FCC is pursuing a policy of stimulating long distance competition in telephony and the enhanced service exemption is just one way of giving IP telephony a chance to gain a foothold in the market. Improving the efficiency of IP telephony may not be sufficient in itself to enable it to compete successfully in the long term with circuit switched voice, voice over ATM and voice over Frame Relay.

However, IP telephony has two overriding advantages that distinguish it from other voice transport technologies. First, it can be integrated with other IP-based services on the user's desktop computer and second, the ubiquity of the Internet enables integration of voice messaging, fax and e-mail in unified messaging applications. These additional services may generate significant revenue for IP telephony providers over and above that generated from the telephony service itself and give them an advantage over telephony providers using other transport technologies. In the circuit switched world, Customer Local Area Signalling Service (CLASS) features generated 3–10 times the revenue originally estimated. Customers, particularly residential customers, are prepared to pay more for call forwarding than for the call that is being forwarded! If similar additional revenues can be obtained by ITSPs, they may more than offset the anticipated loss of the enhanced service provider exemption.

In this section we will describe some examples of IP-based services; unified messaging will be discussed in Section 4.7. Not all of these generate direct revenue for the ITSP, since in some cases the service can be provided by the user themselves using software on their desktop computer. However, in all cases the fact that the service is possible itself stimulates the market for IP telephony.

The examples of IP-based services which follow in general apply to the situation where the caller and the called party *both* use IP telephony from their desktop computers.

### 4.6.1   Instant Messaging

#### 4.6.1.1   Instant Voice Messages

Instant voice messages are the voice equivalent of text-based chat, and are otherwise known as voice chat. There are two major applications.

- Users scan the network to identify whether any of their 'buddies' are on line. If so they can send a voice message to an individual who plays it though their audio playback software.
- Users visit a Web-based chat room in which multiple participants can send voice messages to all other participants.

#### 4.6.1.2   Instant Text Messages

Instant text messages can be used in conjunction with a telephone call. Here are three sample applications.

- If a call is being attempted but the called party is already speaking with someone else, the caller can send an instant message that appears on the called party's screen.
- Another scenario when the called party is busy is for the caller to request the network to monitor the line and inform the caller when the called party becomes free. A message can be flashed on to the caller's screen from the network, in a similar way to the operation of 'Re-dial on Busy' in the circuit switched network.
- In the case of a multiparty voice conference, messages can be sent by the network to each participant's screen indicating who is joining/leaving the conference, plus an identification of who is currently speaking.

### 4.6.2   Web/Telephony Integration

Examples of Web/telephony integration are as follows.

- *First telephony then the Web.* If a customer is calling a company and 'all agents are currently busy', instead of sending the customer music to soothe their impatience, the company can send its home page to the customer's desktop computer. The customer may be able to find the information they want without speaking with an agent.
- *First the Web then telephony.* On the other hand let us suppose a customer has browsed a corporate Web site and needs to speak with an agent. 'Push to Talk' buttons on a Web page can be used to make a connection. Moreover, the call can be made direct to the appropriate department in the company. Instead of having long IVR menus, 'Push 1 for sales, 2 for service, 3 for accounts...', these options can be displayed on a Web page. It is far easier for the customer to browse multiple options on the screen than on an IVR system. When the call is set up to an agent in the company, the agent can be presented with the Web pages that the customer has browsed, so as to get background information on the customer's interests. During the course of the telephone call, the agent can 'push' additional Web pages to the customer's Web browser, e.g. to demonstrate features of the companies products.
- *Web-based call centres.* The architecture of a Web-based call centre providing the functionality just described is shown in Figure 4.30. The Web server can be integrated with the voice gateway and the call control server can be an H.323 gatekeeper or SIP Redirect or Proxy Server. Monitoring customer/agent interactions can lead to redesign of the Web site to make more information accessible to the customer without the need to speak with an agent.
- *International toll-free calls.* When a customer uses a 'Push to Talk' button on a corporate Web page, it can initiate an IP telephony call between the company and the customer anywhere in the world. This is essentially an international toll-free service.
- *Web-based caller ID.* SS7-based caller ID services display the caller's phone number or name to the called party before the called party answers the phone. This can be used by the called party or by their software to determine whether to answer the call, send it to a messaging service or an IVR system, or forward it to a different phone number. Web-based caller ID sends the caller's home page to the called party, thus providing much more information about the caller than can be obtained from their name and phone number. For instance, the Web page can be attached to the SIP 'Invite' message.
- *Advertising-based charging.* An ITSP can set up a Web site for users to place a call from an Internet-attached computer to a PSTN-attached telephone. Voice gateways are required in the destination cities. Some such services charge for calls per minute of usage, while others derive revenue from advertisements displayed on the Web browser of the caller.

### 4.6.3   Document Sharing

One of the most powerful features that the Internet can add to telephony is the ability to share documents, for instance documents relating to the subject matter of the telephone discussion. One participant in the conversation can give the other(s) a slide show using

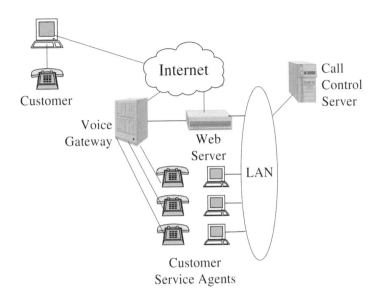

**Figure 4.30**   Web-based call centre: circuit switched voice

graphical images to illustrate what they are saying. Another example is where multiple participants in a call can jointly edit a document that is being discussed. Software to facilitate these applications generally has four modes of operation:

- *Show and tell*. Only the originator of the document can control the document, i.e. which pages are displayed, any changes being made to the document, etc. This is useful when one person is giving a presentation to others.
- *E-mail*. The document is sent as an e-mail attachment to all participants who can manipulate their own copies in whatever way they wish on their own computers.
- *Collaboration or Joint Editing*. The document remains on the originator's computer but any participant in the call can edit and control the document. Contention among multiple participants can be resolved, for instance, by requiring people to click the mouse in order to gain control of the document. There are security concerns when one user can control another user's software, and some corporate firewalls do not allow collaboration with users outside the company.
- *Whiteboard*. Whiteboard is similar to Collaboration or Joint Editing except that in general there is no formal document being jointly edited. Instead, the whiteboard serves as an area where multiple participants can brainstorm and contribute text and graphics to support their ideas. The whiteboard is often not a document that originates on a specific computer (as in the case of Collaboration or Joint Editing) but can be saved by any participant on their own computer.

### 4.6.4   Video Telephony

Ever since AT&T announced a videophone for use on analogue residential access lines in the mid 1970s, vendors have been waiting for the market for videotelephony to take off. The Internet offers an opportunity to conduct desktop videoconferencing. Low cost motion video cameras can be connected to desktop computers together with audio equipment, providing video accompaniment to telephony.

When video is used simply to provide a head and shoulders view of the speaker, many users regard it as an intrusion on their privacy, and not necessary for the other person to understand what they have to say. However, there are specific areas in which video can contribute substantive content to a telephone call. They generally involve a telephone conversation in which the subject matter of the call contains motion video content.

- For remote maintenance and repair of equipment, video can be used to show a remote engineer how equipment is performing, which can help diagnose possible faults. An example application is the airline industry where aircraft are maintained and serviced by airline personnel, who may need to contact the aircraft manufacturer for advice on certain problems. Video can be used to show the manufacturer the problem the airline has encountered.
- In health care, many imaging modalities such as Magnetic Resonance Imaging (MRI) and ultrasound involve motion video. Videotelephony can be used to obtain advice from a remote specialist.
- In education and training, motion video is sometimes useful to show the instructor's gestures and also to show a scientific experiment being conducted by the instructor.

### 4.6.5  Summary

In summary, the integration of telephony with Internet and Web-based services gives an advantage to voice over IP compared to the use of other transport technologies, since IP reaches the user's desktop computer.

'Telephony over the Internet' can be regarded as an old service (telephony) over a new network (the Internet). The Internet provides nothing more than the transport of the voice traffic.

By contrast 'Internet telephony' augments the old telephony functionality by integrating Internet and Web-based services with it. Some of these services offer revenue opportunities to ITSPs.

## 4.7   Unified Messaging

Another service that can be added to telephony, when the Internet is used as the transport network, is unified messaging in which voice messaging, fax and e-mail are integrated together. Figure 4.31 shows a user on the left with a telephone and a computer. In order to access voice mail, the user has two options.

1. Make a conventional call across the PSTN to the voice mail server and navigate through the options available using DTMF signalling.
2. Arrange for the voice mail to be sent as e-mail attachments to his e-mail system. Navigation is now simpler with the use of the keyboard and mouse, and voice mail can be played through the sound card of the computer.

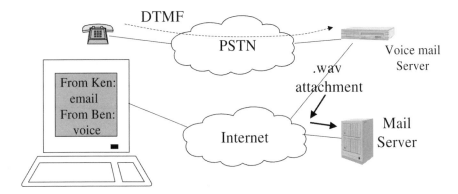

**Figure 4.31**  Unified messaging voice to e-mail

In a similar way, the user can also arrange for faxes to be sent as e-mail attachments instead of being sent to fax equipment attached to the PSTN.

Sending voice mail over an IP network also offers considerable cost savings and can be used in conjunction with a regular telephone handset instead of a computer. Figure 4.32 shows an employee of ABC Inc, who is based in San Diego, making a business trip to the company office in Boston. From the hotel room in Boston, he would normally need to make a PSTN or corporate network call to the San Diego server in order to retrieve his voice mail. Instead he arranges for the voice mail to be sent over the public Internet or the corporate Intranet to the corporate voice mail server in Boston. Now only a local telephone call is needed to receive the voice mail. In this application the user can use a regular hotel room telephone. There is no need to use a computer, unless he also wishes to retrieve e-mail. The reduction in long distance charges for retrieving voice mail on business trips, particularly international ones, has prompted some companies to set up temporary voice mail servers in cities where they do not have a corporate office. For instance, if a large

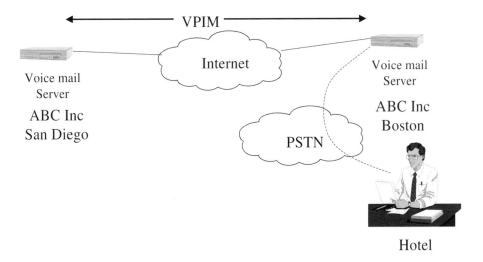

**Figure 4.32**  VPIM enterprise application

number of employees are attending a conference, handling all voice mail on a server in the conference city has been found to be very profitable.

Figure 4.32 indicates that the protocol involved in sending voice mail as e-mail attachments is Voice Profile for Internet Mail (VPIM), which we now describe. This will be followed by a similar review of the fax over the Internet.

### 4.7.1   VPIM

VPIM [8] specifies the way in which a voice message and/or a fax can be attached to an e-mail using Multimedia Internet Mail Extension (MIME). It also specifies formats for phone numbers in E.164 format to be incorporated into e-mail addresses so that the e-mail is addressed to a telephone or fax machine. A server that receives the e-mail with the voice/fax attachment can either

- store the attachment and wait for a user to subsequently retrieve it, which is suited to voice messages;

  or

- forward the attachment to the E.164 number, which is suited to faxes.

Standard E.164 international telephone numbers and/or extension numbers can be incorporated into e-mail addresses in one of the following formats.

- *Private numbering plan.* If ABC Inc has an internal private numbering plan for its phone lines, then a typical address is 4793@abcinc.com.
- *International E.164 number.* An ITSP or other service provider can set up servers to receive VPIM messages over the Internet and forward them over the local Telco network. A typical address for an ITSP's New York server is +12125559876@ny.itsp.com.
- *International E.164 number with extension.* If the E.164 number has internal extensions, the extension number is separated from the E.164 number by a + sign, e.g. +12125559876+4784@ny.itsp.com.

An example of a VPIM message is shown in Figure 4.33. The header of the message with the From, To, cc, and Subject fields corresponds with the format of a standard e-mail message. This immediately gives considerable flexibility compared to regular voice mail in that the message can be sent to multiple recipients with multiple copies to other users. It also allows the use of distribution lists in the To and cc fields.

The MIME extension contains the following fields.

- The first item is the duration of the audio content, so that the recipient knows in advance how long the voice message is, again a feature not available on most voice messaging systems.
- Also there is an indication of which language the voice message is in.
- The spoken name of the sender is included in case the identity of the sender is not clear in the From field.
- Business card style general co-ordinates are included in Versit Card format.
- Finally, we have the voice message coded using 32 Kbps ADPCM and/or the fax coded using Tag Image File Format (TIFF)-F [9].

- From: "Kim User" <12125559876@ny.itsp.com>
- To: 16137639999@nortelnetworks.com
- cc: 12125551234@abcinc.com
- Subject: megaproject
- --------------------MIME CONTENT------------------
  - Audio content duration and language
  - spoken name
  - Versit Card, VCARD
    - job title, phone #, email address
  - Audio message using 32 Kbps ADPCM
  - Fax using TIFF-F image encoding: RFC 2302

**Figure 4.33**   Example VPIM message

.

Replying to and forwarding a VPIM message can be performed with the same ease as for regular e-mail. E-mail features such as 'Reply to all' allow copies of the reply to be sent to all addressees of the original message, which is generally difficult to achieve with most voice messaging systems and fax machines.

VPIM therefore not only allows voice mail and fax to be integrated with e-mail and retrieved using a single piece of software, but also allows the use of e-mail oriented features to provide additional flexibility in the handling of voice mail and fax.

### 4.7.2   Fax over the Internet

A major market for fax over the Internet is international faxes. Because of the considerable time difference on many international calls, callers prefer to send faxes instead of speaking by telephone. In fact, 25% of international calls are faxes. Many countries have fax servers which receive faxes over the Internet and relay them to E.164 numbers within that country. Prior to standardization, proprietary formats were introduced that were specific to an individual country. For instance, to send a fax to a number in Kuwait, it could be attached to an e-mail addressed as follows:

- To: fax@kuwait.net
- Subject: ⟨E.164⟩

The Subject field contains the E.164 address within Kuwait to which the fax should be relayed.

To send a fax to a number in the Netherlands, it could be attached to an e-mail addressed as follows:

- To: ⟨FirstName.LastName.E.164⟩@e-fax.nci.nl

In this case the E.164 address within the Netherlands and the name of the addressee were included in the To field.

The variety of formats chosen by server operators for fax over the Internet has led to the development of standards [10–12] which specify addressing in the format:

- To: ⟨fax=+44716784321@ukfaxserver.com⟩

Faxes are attached as MIME attachments in TIFF. A utility for converting documents to TIFF is available at www.faxaway.com/tiff.html.

Servers relaying faxes to fax machines over the PSTN need to handle situations that are not encountered in the case of voice mail, e.g.

- destination busy;
- no fax tone at destination;
- line error on receiving fax machine.

### 4.7.3  Unified Messaging Summary

The basis for unified messaging is that standards are available for attaching voice mail and fax to e-mail messages. However, unified messaging means different things to different people:

1. receiving voice mail, e-mail and fax within a single e-mail software system on a computer;
2. sending voice mail over an IP network so that it can be retrieved over the local PSTN using a regular telephone;
3. sending fax over an IP network to a server that forwards it over the local PSTN to a regular fax machine.

In 1, it is primarily the *destination* user that has the advantage of 'unification' of messaging since all the messages are available in a single software system on their computer.

In 2 and 3, the 'unification' is more from the viewpoint of the *sender or sending server*, since the fax and voice message are converted to IP format and attached to e-mails before being sent over the IP network. They can be received on regular PSTN-attached fax equipment and telephones.

Of course 1, 2, and 3 can be combined so that, for instance, a fax, originating as hard copy, can be scanned into a computer, attached to an e-mail and sent to another computer where it is printed out, or read on the screen. The variety of options in which 1, 2, and 3 can be combined is the advantage of unified messaging and gives great flexibility to how it can be used.

## 4.8  Case Examples

In this section we describe how voice over IP can be used in the four major markets:

- desktop applications;
- enterprise networking;

- incumbent public carrier networking;
- competitive public carrier networking.

### 4.8.1  Desktop Applications

IP is the prime contender for desktop voice applications, because IP is the major transport technology extending to the desktop. Although ethernet is also present at the desktop, voice over ethernet is of less importance than voice over IP since it is only applicable to LAN applications, and needs to be converted into another technology for longer distance communications.

Our objective is to integrate voice and data traffic onto a single network because we want to make telephony calls or because we want to do multimedia applications, which include voice and data.

The ubiquity of IP at the desktop includes both desktop computers and also IP phones, both of which have significant processing power. Many voice features can therefore be provided from the desktop equipment without interaction with network-based servers. Examples of services that require interaction with network-based equipment when provided over the PSTN to analogue phones, but which can be obtained from software in desktop equipment over an IP network are:

- *Redial on busy*. In the PSTN, for telephone calls, this service is provided by the destination class 5 switch monitoring the called party line, until it becomes free. For fax calls over the PSTN, the source fax machine can keep redialling a busy number until it is free. From the viewpoint of the public carrier, the first of these alternatives is a source of revenue, and the second is an additional cost. In an end-to-end IP network, the calling party equipment can attempt the call repeatedly until successful, resulting in a cost but no additional revenue for the carrier.
- *Calling party ID*. In the IP world, the source IP address is the calling party ID. Since IP equipment knows its IP address, the source can provide its own IP address to the called party in the call set-up messages. At the destination, filtering incoming calls according to calling party ID can be performed in software in the desktop equipment, forwarding some calls to other numbers, sending others to voice mail and alerting the user for those they need to answer right away.
- *Call waiting*. Since an analogue POTS line cannot carry two calls at the same time, in the PSTN, call waiting is a service provided by the destination class 5 switch. In an IP network, a user can simultaneously receive packets from multiple sources. Software in desktop equipment can be used to put one call on hold while speaking on another call.
- *Unified messaging*. In the PSTN, servers are required to forward voice messages and faxes as e-mail attachments to a customer's e-mail address. In an IP network, desktop equipment can route messages in a flexible manner determined by the user.

Section 4.6 describes many features available from an IP desktop telephony application that simply are not possible with the PSTN.

There remain some services for which an IP desktop telephony user still requires network-based software, thus providing revenue opportunities for the carrier.

- *Billing*. Flexible billing options, such as charging calls to a credit card or other account, require authorization from a network server.

- *Call forwarding.* In the PSTN, call forwarding is typically implemented in the destination class 5 switch. In an IP network, SIP provides very flexible call forwarding, including different forwarding options according to date and time. These require interaction between users and network-based SIP servers. A company can set up SIP servers on its corporate Intranet for forwarding internal calls, and for off-net call forwarding, interaction with public network-based SIP servers is required.

### 4.8.2   Enterprise Network

Prior to the introduction of packet voice, enterprise voice networks were based on leased T1/E1 lines as shown in Figure 4.34. Each PBX is connected to other PBXs via leased lines, but is not necessarily connected to *every* other PBX. Traffic from San Francisco to New York in Figure 4.34 needs to go via Chicago or Denver. The resulting 'tandem' traffic through Chicago and Denver results in additional processing in the PBXs in the corporate customer sites in those cities. PBXs can compress voice but do not implement silence removal, since the traffic is transported in T1/E1 time slots using constant bit rate circuit switching. Data traffic typically follows a similar but separate leased line network.

Figure 4.35 shows how the voice can be integrated with the data traffic in IADs at the customer premises and sent over a common network. The IAD removes silence from the voice, thus saving about 60% of the voice bandwidth so that it may be possible to transport the corporate data traffic in the saved bandwidth. If so we essentially have 'free data' since both voice and data are transported on the same network that had previously been transporting voice only. In this scenario, the tandem traffic is a burden to the IAD, but not to the PBX.

An alternative strategy for integrated corporate communications over IP is shown in Figure 4.36, where the company subscribes to a public IP service instead of sending its IP traffic over leased lines. In this case, the IP network is managed by a public carrier so as to provide QoS to the voice and data traffic. A SLA specifies the QoS and other parameters as described in Section 2.1. The company pays the service provider a price dependent on the total traffic entering and leaving the network in each city, and may also set up a Virtual Private Network with specified bandwidth and QoS on each connection between each pair of corporate sites. In this configuration the customer equipment is relieved of all tandem traffic.

The business case for choosing among these three alternatives depends on:

- *the cost of the IAD* which may be purchased by the customer as a capital expenditure or

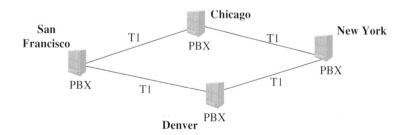

**Figure 4.34**   Enterprise voice network based on leased lines

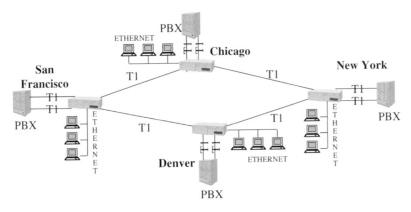

**Figure 4.35** Enterprise network: data and voice sharing leased lines

it may be leased to the customer by the carrier. Leasing is a common option in the scenario in Figure 4.36, since the IAD needs to interwork with the network equipment. Leasing allows the carrier to select equipment for the customer premises which they know to be compatible with their own network equipment. The IAD would come from the same vendor as the network equipment or from a vendor with which interoperability tests have been conducted.

- *the exact distribution of traffic between each pair of sites.* In the leased line options in Figures 4.34 and 4.35, the customer is restricted to the bandwidths of leased lines that are available from the carrier, typically ½ T1/E1 or full T1/E1. If the customer's traffic is ¾ T1/E1, they need to use and pay for a full T1/E1. The scenario in Figure 4.36 has the advantage of greater flexibility in the bandwidth on the Virtual Private Network between cities. Carriers offer bandwidths that can be more closely matched to the customer's traffic profile. This can result in cost saving to the customer since they are not purchasing unused bandwidth.

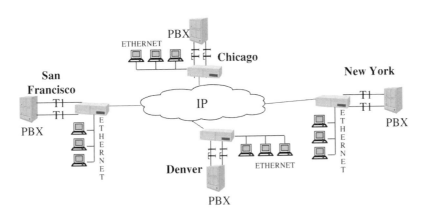

**Figure 4.36** Enterprise network: data and voice over an IP service

### 4.8.3   Incumbent Public Carrier Networking

An incumbent public carrier can use voice over IP within its own network. This is usually done in the backbone of the network with voice traffic brought to the backbone by a circuit switched edge network composed of class 5 and class 4 (tandem) switches. The increase in circuit switched traffic as a result of (a) reduced tariffs increasing demand for phone calls and (b) long holding times on dial-up modem access to the public Internet has resulted in carriers experiencing the problem of 'tandem exhaust' where the tandem switch runs out of capacity. Excess traffic can be offloaded onto an existing managed IP backbone via a MG, thus alleviating the congestion. The traffic still comes through the class 5 switch from which it accesses all the software needed to provide voice call features. Some vendors of class 5 switches implement a packet interface on the switch itself which provides the functionality of a MG.

In the case where the class 5 switch is hitting its capacity limits, voice traffic can be offloaded onto a managed IP backbone network via a MG, thus bypassing the class 5 switch. In this case voice features need to be provided by a feature server attached to the IP network.

### 4.8.4   Competitive Public Carrier Networking

A competitive public carrier building a network in a greenfield situation is fundamentally different from an incumbent public carrier if it does not have an installed base of class 5 circuit switches. In order to transport voice traffic, the competitive carrier needs to address the question of providing voice features in addition to the transport of the voice traffic itself. There are basically two approaches to this problem.

- The competitive carrier can focus its marketing on medium to large business customers who have PBXs at their corporate sites. All the voice features are obtained from the PBXs and do not need to be provided by the network operator. The PBX traffic enters the public network at a MG which packetizes it for transport over a managed IP backbone.
- The competitive carrier can implement feature servers attached to the managed IP network. The customer's voice traffic enters the network through MGs which use messaging to obtain the features from the feature servers.

## 4.9   Summary

It is the last few moments before sunset that prove that white light is composed of a spectrum of colours. High in the Sierra Nevada, surrounded by snow, with fluffy cumulus above, sunsets can be magnificent. Neither snow nor cloud is white. The diamond snow crystals underfoot flashing with the last rays of the sun give way to amorphous pink and ruby as one looks further away, until the ruby darkens into a mauve silhouette of the distant mountains. The clouds above change every moment through a thousand shades of apricot and amber, and if a cloud should attempt to block the sum, that fierce orb will, as if in retaliation, fire orange daggers of light past its edges. Some dark clouds do not have silver linings but golden ones instead.

Like white light, IP is a many splendoured thing. It is a backbone technology and an edge technology. It can multiplex many voice calls providing trunking over the backbone and it can interwork with desktop software to provide user-defined features on individual calls. It can marry voice functionality with data functionality such as the Web. It can look like a PBX or Centrex, and above all, for voice, it provides two alternatives for QoS: RSVP and DiffServ.

## References

1. Flanagan, P. (1999) Next Generation Telcos – Preparing for Market Impact. *Business Communications Review*, 23–28. Duke, J. (1999) VoIP Market – the Game Has Begun. *Business Communications Review*, 12–13.
2. www.feldmancommunications.com.
3. RTP Multiplexing: draft-ietf-avt-aggregation-00.txt.
4. RTP Multiplexing: draft-ietf-avt-mux-rtp-00.txt.
5. IETF RFC2508 (1999) Compressing IP/UDP/RTP Headers for Low-Speed Serial Links.
6. ITU-T Recommendation H.323 (1999) Packet-Based Multimedia Communications Systems.
7. IETF RFC2543 (1999) SIP: Session Initiation Protocol.
8. IETF RFC2421 (1998) VPIM, Voice Profile for Internet Mail.
9. IETF RFC2302 (1998) Tag Image File Format (TIFF) – image/tiff MIME Sub-Type Registration.
10. ITU-T Recommendation T.37 (1998) Procedures for the Transfer of Facsimile Data Via Store-and-Forward on the Internet.
11. IETF RFC2301 (1998) File Format for Internet Fax.
12. IETF RFC2306 (1998) Tag Image File Format (TIFF) – F Profile for Facsimile.

# 5

# Voice Over ATM

## 5.1 ATM Market Acceptance

ATM was developed in the late 1980s, at a time when the functionality that could be implemented in silicon was far more limited than it is today. Because of these limitations, ATM was designed to have fixed length data units, in order to facilitate hardware-based switching, since less operations are required to switch fixed length data units than variable length data units. Fixed length data units are unusual in an industry which uses variable length for most transport technologies, e.g. ethernet, TCP, IP, and Frame Relay. To distinguish ATM data units from the data units used in other technologies, they are referred to as 'cells'.

During the first half of the 1990s, hardware-based switching resulted in ATM switches having higher throughput than switches for other technologies. ATM was therefore deployed in the backbone of many carrier networks, including ISPs and telephone companies. LAN backbones also saw the introduction of ATM.

Today variable length packets can be switched in silicon as fast as fixed length cells, so that ATM has lost its initial competitive advantage over other technologies. There are two main reasons why ATM is important today.

- First, the deployment of ATM in all major network backbones means that it is an established part of the telecommunications landscape. Although other technologies could equally well be used in a network backbone being built today, there is little business reason to remove ATM equipment in order to replace it by another technology.
- However, ATM is not simply a legacy technology that we have to contend with until it dies. The intense industry interest in ATM during the 1990s resulted in extensive standardization of many useful features associated with ATM. Of primary interest for voice transport are the sophisticated trunking and QoS standards including QoS-based routing.

## 5.2 ATM Technology Primer

### 5.2.1 Header Functions

ATM cells have a 48 byte payload and a 5 byte header. Figure 5.1 illustrates the header

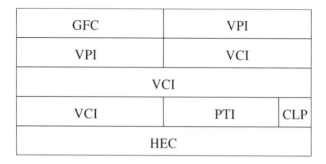

**Figure 5.1**  ATM cell header

fields with 5 bytes vertically and 8 bits horizontally. The header fields provide the following functions:

- *Generic Flow Control (GFC)*. GFC resolves contention among multiple users on a
  single interface. GFC was designed for multiple desktop computers sharing a single
  switch port, and allows limits to be placed on the percentage of the port bandwidth that
  can be used by each desktop computer.
- *Virtual Path/Channel Identifier (VPI/VCI)*. ATM virtual circuits are uniquely identified
  on any given link in the network by a VPI/VCI. The use of two identifiers provides two
  levels of multiplexing in an ATM network. Figure 5.2 illustrates how virtual channels
  can be aggregated into virtual paths. If a virtual channel is used for a single telephone
  call, multiple calls can be transported across the ATM backbone in a single virtual path,
  greatly simplifying network management in the backbone. Only one entity shows up in
  the network management system, namely the virtual path, instead of a large number of
  individual virtual channels: one for each phone call. In this chapter we use the term
  ''virtual circuit'' to mean either a virtual channel or a virtual path.
- *Payload Type Indicator (PTI)*. PTI distinguishes between different types of payload,
  e.g. network management and user traffic. In the case of user traffic a bit in the PTI can
  be used to indicate that the cell passed through a congested switch. In the case of user
  ATM Adaptation Layer Type 5 (AAL5) traffic a bit can be set to indicate the last cell of
  a higher level data unit.

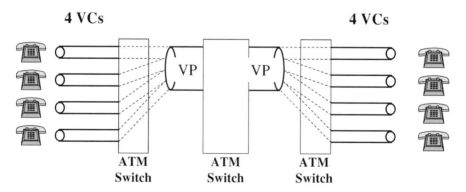

**Figure 5.2**  Two levels of multiplexing

- *Cell Loss Priority (CLP)*. CLP is used to distinguish between low priority and regular priority cells within a *single* virtual channel. A major use of CLP is that it allows a network operator to downgrade the priority of user traffic if it exceeds the cell rate agreed in a traffic contract between the user and the network operator. CLP does not deal with the fact that some virtual channels have higher priority than others, which is handled by proprietary mechanisms within switches.
- *Header Error Control (HEC)*. HEC provides a bit error check on the other 4 bytes of the header, which can be use for error detection and optionally for error correction.

The ATM header may well be familiar to most readers of this book. One criticism of ATM is that the 5 byte header results in about 11% overhead for a 53 byte cell. This is sometimes referred to as the 'cell tax'. For voice transport, as we saw in Section 4.3, the 'tax' can be much higher for IP than for ATM. The 5 byte ATM header is much shorter than the 20 byte IPv4 header. One point that is important to emphasize about ATM is that the header functions are simple. The whole idea of ATM is to strip away the functionality out of the network core, so that switches can concentrate on switching the cells through the network with a high throughput. The minimum functionality required in the network core was designed for early implementation in hardware. More processing-intensive functionality, such as packing customer traffic into ATM payloads and interworking with other transport technologies, is implemented at the ATM network edge.

## 5.2.2   QoS

ATM provides QoS in two respects. One is the accuracy of the communication, and the other is the end-to-end delay across the ATM network. There are a number of measures of accuracy and delay which are useful in monitoring the performance of the network, but there are only three which the customer can specify. When the customer is specifying a connection that they want to set up across an ATM network, they can request a certain Cell Loss Ratio (CLR), a certain Cell Transfer Delay (CTD) and/or a certain Cell Delay Variation (CDV). Those are the three key parameters for the network to be able to control, because those are the three parameters which the customer can request from the network. An example of a QoS parameter that can be monitored by network equipment but cannot be requested by a customer is cell error ratio: the proportion of cells containing bit errors in their payloads. The definition of the three parameters that can be requested by the customer is as follows:

- The CLR is the proportion of cells on a specific virtual circuit that are lost while traversing the ATM network. Loss of information is very important in most data applications, and less important in most voice applications, because the human ear can maintain intelligibility of human speech even if a few milliseconds are lost. However, it is important to distinguish between cell loss and information loss. Cell loss refers to the loss of an ATM cell in a network. Information loss refers to the loss of information being processed by a user's software application, e.g. a spreadsheet in the case of data and speech playout software in the case of voice. Although the data application is sensitive to information loss, there are other error control mechanisms besides CLR that can be used in the case of errors in a data communication, e.g. the TCP mechanisms for retransmission. CLR can nevertheless be used for data traffic, for instance to reduce the number of TCP retransmissions, because when TCP retransmits, it retransmits the entire TCP packet, which can be several thousand bytes long. This retransmission may have been caused by the loss of only one ATM cell within the TCP packet. So TCP can retransmit data traffic but generally speaking we set the CLR so as to reduce the TCP retransmissions. TCP cannot be used for interactive voice applications since there is not time for the retransmissions. The concealment mechanisms described in Section 2.2.3 and CLR are therefore the main error control mechanisms for interactive voice. The importance of CLR depends on how compressed the voice is. An ATM cell containing 64 Kbps voice represents only about 6 ms of voice. However, if the voice is compressed to 8 Kbps, the cell contains about 48 ms of voice. Concealment mechanisms become less effective when longer voice packets are lost, and therefore CLR becomes very important when compressed voice is transported over a single ATM virtual circuit. When multiple compressed voice calls are carried over a single ATM virtual circuit, e.g. using AAL1 or AAL2, the loss of an ATM cell may represent the loss of a relatively small amount of voice from each voice call, which can be handled by concealment methods.

- With CTD, each cell that traverses an ATM network takes a different amount of time. The buffer fill levels encountered by each cell in each switch are continuously varying, resulting in variable delay. Within a specific virtual circuit, there is a distribution of end-to-end delay (Figure 5.3). No cell can traverse the network in less than a certain delay determined by the speed of light. The majority of cells traverse the network with a medium amount of delay. Finally, there are some exceptional cells that encounter unusual congestion. The delay distribution therefore has a long tail, due to these unlucky cells. When a carrier guarantees to deliver a customer's cells within a certain period of time, that guarantee does not apply to the tail of the distribution. Instead CTD is defined statistically as a percentile of the delay distribution. We measure the CTD in milliseconds. Thus, a customer requesting a voice connection may request, for instance, CTD < 100 ms on 99.9% of the cells. CTD is therefore not an absolute guarantee on *all* cells but a guarantee on a certain percentage of cells which is specified at the time the call is established. It is important to note that CTD measures the delay end to end across

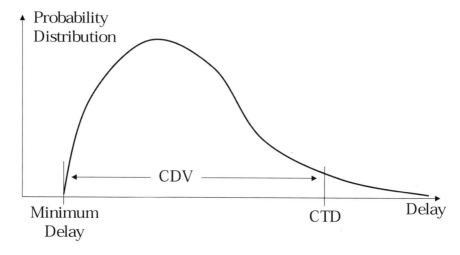

**Figure 5.3**   ATM delay QoS

an ATM network. The delay perceived by the user may be greater than this if the traffic has to traverse other networks outside the ATM portion of the network and also due to processing delays at ingress to and egress from the ATM network as described in detail in Section 2.3.

- CDV measures the extent of variability in the delay among cells within a single virtual circuit. We measure the CDV in milliseconds. As with CTD, it does not include the variability due to the exceptional cells. CDV is defined statistically as the difference between the maximum delay encountered by a given percentile of cells and the minimum delay with which it is possible to traverse the network (Figure 5.3).

The extent of delay and delay variability is dependent upon the traffic conditions in the network. A lot of bursty data traffic in the network impacts the CDV of other traffic. All cells traverse the switching fabric of ATM switches in terms of microseconds rather that milliseconds. But once a cell has got through the switching fabric, it has to wait in the output buffer for an amount of time which depends upon the volume of other traffic in the buffer. When a customer sets up a voice call requesting a specific CTD and CDV, each switch examines its current traffic load and attempts to assign a priority level (internal to the switch) which allows the new voice call to obtain the requested QoS, without adversely affecting the QoS of existing traffic flows. If this is possible the call is accepted, if not it is rejected.

### 5.2.3   Service Categories

There are six main categories of service in ATM.

- Two real time categories:
  - Constant Bit Rate (CBR);
  - Real Time Variable Bit Rate (rt-VBR).

- Four non-real time categories:
  - Non-Real Time Variable Bit Rate (nrt-VBR);
  - Unspecified Bit Rate (UBR);
  - Available Bit Rate (ABR);
  - Guaranteed Frame Rate (GFR).

The main difference between the non-real time categories and the real time categories is that a user of a real time category can request CLR, CTD and CDV from the network when setting up the connection. However, the non-real time users can only request CLR, except for UBR, which as its name suggests does not receive any guarantee from the network about QoS or anything else. The service category is specified by the customer at the time a connection is established along with the QoS and the cell rate. Interactive voice applications such as telephony require a real time service category. Other voice traffic can use non-real time service categories.

For VBR service categories, two cell rates are specified: a peak cell rate (PCR) which is designed to allow occasional bursts in traffic and a Sustainable Cell Rate (SCR) which the customer can maintain for more extended periods of time.

For ABR, the network guarantees a minimum cell rate, but if the customer is offering traffic at a higher cell rate than the minimum, and the network gets congested, the network may require the user to slacken off with their traffic and reduce down to a lower cell rate. If the network doesn't get congested, then the user can increase to a peak rate and at any time the user is using a cell rate between the minimum and the peak. The important distinguishing feature about ABR is that the network can tell the user to reduce their offered traffic in the case of network congestion. The customer is therefore assisting the carrier in congestion control, which is important as ATM network traffic increases in volume and also becomes more bursty.

For GFR, the network guarantees to carry the customer's traffic in complete higher level data units, e.g. IP packets, up to a certain cell rate. Above that rate the network may, under congestion conditions, discard individual cells from within a higher level data unit.

### 5.2.4   SVC and Permanent Virtual Circuits

A SVC is a connection that is set up on demand by the user. Standardized signalling messages are sent from the user to the network, User-to-Network Interface (UNI) signalling [1], and across the network from switch to switch [2], PNNI. The signalling messages contain information about the ATM source and destination addresses, the cell rate required, the service category, the QoS and the ATM Adaptation Layer (AAL).

A Permanent Virtual Circuit (PVC) is set up using the network management system, which may involve proprietary messaging, and/or may use PNNI.

*Individual* voice calls typically need to be set up on demand and therefore require SVC capability in the network. *Multiplexes* of phone calls, for instance a T1/E1 or fractional T1/E1 interconnecting PBXs among the sites of a corporation, are required on a more permanent basis and are typically set up as PVCs.

### 5.2.5   ATM Signalling

The basic functionality of UNI signalling is provided by UNI 3.1, which covers basic call establishment for certain service classes. UNI 4.1 adds the following capabilities:

- *Signalling messages from multiple users on a single interface.* This is useful, for instance, in the case of multiple users, each with a DSL access line. All the DSL traffic is aggregated by a Digital Subscriber Line Access Multiplexer (DSLAM) into a single interface (e.g. OC-3) to the ATM switch.
- *Multiparty conferencing.* This allows users to set up conference calls, and multicasts, including group-addressing options.
- *Negotiation of QoS.* Suppose a user requests a certain level of QoS from the network which the network is unable to provide. Instead of rejecting the call, the network can offer the user a lower grade of QoS. If the user accepts, the connection is made at the lower grade of QoS.
- *ABR traffic class.* ABR traffic class was not available at the time UNI 3.1 signalling was developed. UNI 4.1 signalling allows the communication of the many extra parameters required to establish an ABR connection, e.g. the rate at which the user can increase/decrease their offered traffic in different circumstances.

PNNI is a very powerful protocol including not only signalling but also routing. It divides the network hierarchically into switches and various levels of sub-networks and networks. The routing protocol allows each switch, sub-network and network to provide information on its performance to each other switch, sub-network and network, including:

- architecture changes, such as a new switch or new interface being deployed;
- bandwidth availability on a given physical interface calculated as interface speed − total SCR of all existing connections − margin for existing burstiness;
- performance including:

    - CTD;
    - CDV;
    - CLR.

When a switch needs to establish a new connection, the above information is used to identify bottlenecks in the network and plan a path around them. Specifically, the source switch can plan a route for the new connection across the network to its destination, in such a way as to deliver the bandwidth and QoS required by the user. This is very important for voice since the connection can be routed in such a way as to deliver voice cells to the destination within the required end-to-end delay and delay variation. The result of this routing decision is a Designated Transit List (DTL) specifying the intermediate transit points (switches and subnetworks) in the network through which the connection is to be routed. This is known as 'source routing' since the routing decision as to the path the connection is to take is made by the source switch.

In this respect PNNI routing is different from the routing used in other networks. In the circuit switched PSTN, each switch makes a decision as to the 'next hop' to the next switch only. Then that switch makes a similar 'next hop' routing decision until the connection eventually reaches its destination. The same is true of Internet routing protocols such as Open Shortest Path First (OSPF): each router determines the 'next hop' to the next router. The PSTN and the Internet therefore distribute the routing decision over all switches/routers in the network, whereas PNNI centralizes the decision in the source switch. The main advantage of source routing is that once the source routing decision is made, the call is set up fast since no routing decisions are required from all the other switches between source and destination.

In addition to the routing protocol, PNNI also includes a signalling protocol which is used after the routing decision is made by the source switch. The signalling messages contain the DTL and follow the DTL path through the network establishing the connection via the specified intermediate switches. There is inevitably a delay between the routing messages arriving at the source switch stating the bandwidth and QoS that are available on a certain link, and the signalling messages arriving on that link to set up a connection. It is possible therefore that the signalling messages may need to set up a connection on a link where bandwidth or QoS are inadequate. In this case a 'crank back' procedure is used in which the signalling messages crank back to a previous network node which calculates an alternate route from the routing information that it has received. This results in a revised DTL being inserted into the signalling message so that it can try again via a new route.

The ATM signalling used to set up SVCs may need to be interworked with non-ATM signalling. There are three scenarios for this:

- *Access to SS7 databases.* An ATM customer may wish to dial a toll-free number and get the routing for that number off a database in the SS7 network. For ATM calls to access those toll-free number databases, we need interworking between the ATM signalling and the SS7 signalling in order to get to the appropriate database in the SS7 network. Another example is charging a call to a calling card or credit card. There is no point in duplicating the credit validation databases on the ATM network when they are already available on the SS7 network.

- *Call set-up over an ATM and circuit network.* A caller on a PSTN line may wish to call a customer who is subscribed to ATM. The signalling messages generated by the circuit switched PSTN traverse the SS7 network until they reach the interface with the ATM network where they are translated into ATM signalling messages.

- *Call set-up involving MEGACO, MGCP, or H.323.* Chapter 3 describes how MEGACO can be used in conjunction with an ATM backbone network to set up a call. The MEGACO messages from the MGC instruct the MGs of the source and destination addresses of the call across the ATM network and then ATM signalling is used to establish an SVC. The same interworking is required with MGCP and H.323.

### 5.2.6   Voice Options

Interactive telephony, with a person at either end of a voice call, is one of the most demanding types of traffic for an ATM network to transport. It requires a tight end-to-end delay, and delay variation, with a specification of CTD and CDV, and is typically transported using CBR or rt-VBR service categories.

Other types of voice such as fax, modem data and DTMF are less sensitive to delay and can therefore be transported by nrt-VBR or ABR service categories with no CTD or CDV specified. However, it is not always possible to know in advance whether a connection is to be used for interactive speech or another voice application.

Once a connection has been established with a certain service category and QoS, it cannot be changed during the course of the call. If an interactive telephony call reverted to fax during the call, the fax would therefore be sent using the more stringent specification. If interactive speech was required during a nrt-VBR fax call, the speech delay may not be acceptable.

## 5.3   Adaptation Options

In ATM as much functionality as possible is implemented at the network edge, including the ATM adaptation layer which performs various functions including how voice is put into the ATM payload. Adaptation is not a function on ATM backbone switches. It is a function at the edge of the network, implemented on an interface card on an edge switch, or on a desktop workstation interface card. The basic function of the adaptation layer is to put non-ATM traffic into ATM format.

The different adaptation types which can be used for voice are listed in Figure 5.4 and we now describe each in turn. In Section 5.4 we describe how these adaptation options can be used by IWFs in a trunking environment to carry voice over an ATM backbone.

### 5.3.1   AAL1

The first is AAL1 specified in [3]. It is designed to transport real time CBR traffic. It uses one byte of payload out of the 48 so there are 47 left for voice (Figure 5.5). In that one byte

| AAL Type | Functionality |
|----------|---------------|
|          |               |
| 1/voice | 64 Kbps single voice call |
| 1/UDT** | Transport of a continuous bitstream of CBR traffic<br>Clock recovery |
| 1/SDT* | Transport of CBR traffic in blocks of constant length<br>Clock recovery |
| 2 | Packing multiple voice circuits on a single ATM VC |
| 5 | Bit error control<br>Segmentation and reassembly of user data units over 40 bytes |

*SDT:    Structured data transfer
**UDT:  Unstructured data transfer

**Figure 5.4**    ATM adaptation layer. *SDT, Structured Data Transfer; **UDT, Unstructured Data Transfer

there is a sequence number and a single CSI bit, which is used for clock synchronization. In some voice applications the equipment at either end has clocks for which there is an end-to-end timing requirement to synchronize the clocks. The sequence count is also very important for voice because it allows us to put sequence numbers onto our voice cells, up to a maximum of 8, since the sequence count is 3 bits. The purpose of sequence numbers as we saw in Section 2.2 is that they enable the destination to determine whether a cell was lost en route across the ATM network and to take appropriate action to compensate for the lost voice information.

Clock synchronization is required in certain voice applications and not in others. In the case of a single phone call between desktop equipment that does not use clocks, there is no need for clock synchronization. The primary circumstance when clock synchronization is required is when we are taking an entire T1/E1 containing multiple voice calls and sending it from one PBX to another PBX. The clocks in the PBXs need to be synchronized, so that the rate of sending and receiving T1/E1 frames is the same.

There are several options for clock synchronization. The CSI bit *can* be used for clock

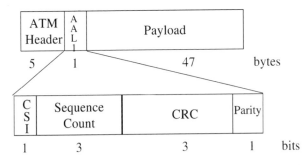

**Figure 5.5**    AAL1. CSI, Convergence Sublayer Indication; CRC, Cyclical Redundancy Check

synchronization but it is also possible to do clock synchronization without it. There are three methods of clock synchronization without the CSI bit and one method that uses it.

- The simplest option for clock synchronization without using the CSI bit is to lock the clocks at both ends onto a common clock. If the ATM network has got one single network clock, the source and the destination can get loop timing off that common clock.
- Even if the network doesn't have a single clock, the edge devices can derive a common clock off the Global Positioning Satellites (GPS) at the expense of GPS receivers in each edge device.
- The third way that we can do clock synchronization without using the AAL1 CSI bit is called 'adaptive clocking' (Figure 5.6). It involves monitoring the buffer in the receiving equipment. If the buffer is getting too full, we've got to speed up our clock so that it clocks the packets out of the buffer faster. This prevents the buffer overflowing, and also automatically adjusts the receive clock to match the transmit clock. Similarly, if the buffer gets too low, we don't want to run out of cells so we slow down the receive clock. This is called adaptive clocking because the receive clock is adapting to the speed of the sending clock. When using adaptive clocking an important question is which end adapts to which? At call set-up time, the calling and called parties use signalling messages to indicate whether they are using adaptive clocking. Suppose the originator of the call says, 'I don't have access to the network clock. So I'm going to use adaptive timing.' Now the important thing here is that if we're using adaptive clocking, we're monitoring the fill level of the buffer, so that our clock is adapting to the clock at the other end. It is impossible for both ends to use adaptive timing because otherwise they are both trying to adapt to each other, and there's no common stable clock anywhere. So if the originator of the call says that it is going to use adaptive timing, then the called party must not use adaptive timing. They must use the network clock or their own clock. The basic principle is not to use adaptive timing at both ends.

These three methods are available in AAL1, but they do not use the AAL1 overhead.

The next method does need overhead, namely the CSI bit, and this is called the Synchronous Residual Time Stamp (SRTS) method. It is illustrated in Figure 5.7 and assumes the existence of three clocks: a network clock, a source clock, and a destination clock. Our aim is to make the destination clock run at the same speed as the source clock. So at the source we measure the difference between our source clock and the network

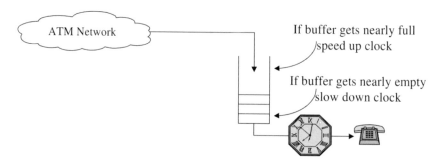

**Figure 5.6**  AAL1 adaptive clocking

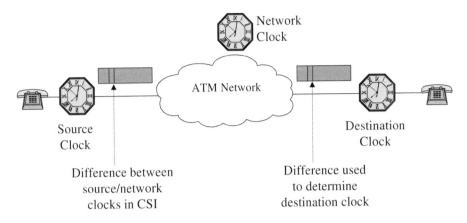

**Figure 5.7**   AAL1: clock synchronization using SRTS

clock. Whatever the difference is, we put it into this CSI AAL1 overhead bit, coded in a way that uses the single CSI bit in several consecutive ATM cells. Then, at the destination, we read the network clock, we read the difference from the CSI bits, and we can calculate what our destination clock should be. So we are taking the differences between the source and the network clock and using them to construct the speed at which the destination clock should run.

AAL1 has an option known as partial fill (Figure 5.8). We may decide to use only part of the 47 bytes of payload in AAL1 in order to reduce the cell assembly delay. G.711 64 Kbps voice generates 1 byte of voice code every 125 µs. It therefore takes $47 \times 0.125 = 5.875$ ms to fill the 47 byte AAL1 payload. Cell assembly delay is more significant with compressed voice. At 8 Kbps, the cell assembly delay is $64/8 = 8$ times as long as it is for 64 Kbps voice, i.e. 47 ms. If we need to transport compressed voice and reduce the cell assembly delay, then we can partially fill the payload. We have $M$ bytes that are used and $47 - M$ bytes which are not used and we can select $M$ to give an acceptable cell assembly delay. We are wasting bandwidth by having an unused part of the cell so it makes sense to use partial fill in parts of our network where bandwidth is low cost, such as in the LAN environment.

Forward Error Correction (FEC) is an option which is available in AAL1. Error correction is based on a block of information, 47 bytes deep, the length of the AAL1 payload, and 124 bytes across, i.e. it corresponds to 124 AAL1 payloads. The error correction code is another four AAL1 payloads making a total of 128. FEC is used when we don't mind about delay, because we're going to get quite a bit of delay accumulating 124 ATM cells and then doing the FEC calculation. FEC is non-standard for single voice calls, according to the ITU. We can use it for circuit emulation using SDT. An example of delay-insensitive

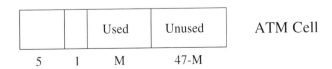

**Figure 5.8**   AAL1: partial fill

traffic that can benefit from FEC is video retrieval or video on demand where we're downloading video from a video server to a user.

We have just described the main features of AAL1 in general. Other features are described below at the point where they are used. There are AAL1s of different flavours, i.e. standardized implementations with certain combinations of AAL1 features. We can regard the features of AAL1 which are described above like a tool kit. We can pick out the tools that we need for different types of traffic. Voice is one type of AAL1 traffic, video is another. So let's pick out the tools which we need for voice. We now describe the three AAL1 flavours that are relevant to voice transport.

### 5.3.1.1   AAL1/Voice

The first flavour of AAL1 we will call AAL1/voice. It is intended to transport *individual* CBR phone calls and therefore needs to be implemented at each end of each phone call. This requires a low cost version of AAL1, and AAL1/voice is a streamlined version of AAL1 in which not all options can be implemented. This enables implementations to reduce cost.

- The first aspect of streamlining is no partial fill. Partial fill requires extra electronics to figure out how full the cells are. So to reduce cost we decide not to do it.
- The second simplification is no cell misinsertion detection. Cell misinsertion detection prevents the network operator from delivering an ATM cell to the wrong destination. We decide not to implement cell misinsertion detection since one cell from a voice conversation does not contain much information and is not worth an extra implementation cost.
- Cell loss detection should be implemented using a simple algorithm to check the sequence numbers, that does not introduce any more delay.
- The SRTS and the adaptive clock mechanisms are not implemented.

AAL1/voice is primarily used for the transport of single G.711 64 Kbps voice calls, and can be regarded as a stop-gap measure prior to the implementation of AAL2 for compressed voice. AAL2 is more bandwidth-efficient as a result of voice compression, but was implemented several years after AAL1. The next two flavours of AAL1 transport several voice calls in a multiplex for which voice compression is inappropriate and are therefore the mainstream applications of AAL1.

### 5.3.1.2   AAL1/UDT

The next flavour of AAL1 is UDT, which is used when we need to transport an entire T1 or E1 frame. The individual time slots within the T1/E1 frame may contain voice and/or 64 Kbps circuit data traffic. AAL1/UDT transports the entire T1/E1 frame independent of its contents.

The T1 frame for instance consists of 24 bytes and 1 framing bit (a total of 193 bits), and we regard it as a continuous stream of bits that are inserted into the AAL1 payload without any reference to the T1 framing (Figure 5.9). We have 6 bytes of overhead: 5 for the ATM header and 1 for AAL1. Then a T1 frame of 193 bits is inserted into the AAL1 payload followed by another T1 frame. We end up with the 23rd byte of the second T1 frame overlapping two ATM cells: with 6 bits from it in one ATM cell and 2 more bits in the

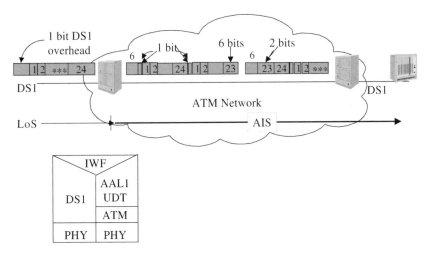

**Figure 5.9**   AAL1/UDT. AIS, Alarm Indication Signal; LoS, Loss of Signal

following ATM cell. There is therefore no byte alignment on the mapping of the E1s or T1s into the ATM cells. At the ATM level, we just view the traffic as being a continuous stream of bits without regard for the structure of those bits. That is why it is called 'unstructured' data transfer. The structure is recovered by the downstream E1/T1 equipment that the traffic encounters after it has exited the ATM network.

AAL1/UDT is usually implemented in Inter Working Functions (IWFs) that interface between a circuit switched network and an ATM network. It is important to note that an IWF is a function, not a piece of equipment, i.e. it represents some processing that needs to be done. Usually we represent the IWF diagrammatically as a box in the middle of an interface, but the function can be implemented either on the equipment at one end of that interface or at the other end. The IWF is typically implemented in a T1/E1 interface card in an ATM switch. It maps the T1/E1 frames into ATM cells as described above, and it also provides some network management functions. For instance, suppose there is a LoS on the DS1 upstream of the ATM network. The next piece of downstream T1 equipment will detect that fault and inform the T1 network management system. However, the next piece of DS1 equipment is at the other side of the ATM network, so that when it reports a fault, it will not be able to locate the fault: it could be upstream of the ATM network, within the ATM network or downstream of the ATM network. UDT *by itself* doesn't help with locating faults on DS1 equipment. That is where standardized interworking comes in. With IWFs, the LoS is recognized by the ATM equipment and converted into an AIS which is fed downstream through the ATM network. Using this information, the location of the fault can be identified more precisely. So we usually implement UDT with the IWF which has this fault management capability.

### 5.3.1.3   AAL1/SDT

SDT is used if we want to maintain knowledge of the structure of the data inside the ATM network. For instance, one example of a data structure is a video conference using H.261 coding at six multiples of 64 Kbps/s. Another example is a fractional T1 service, e.g. 12

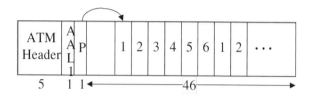

**Figure 5.10**    AAL1/SDT. P, pointer indicating start of 6 × 64 Kbps structure

multiples of 64 Kbps/s. These structures must be maintained across the ATM network and delivered to the downstream equipment. AAL1/SDT transports multiples of 64 Kbps independent of whether the individual 64 Kbps time slots represent voice, 64 Kbps circuit data, or CBR video.

We align the bytes from our traffic into the bytes of the ATM payload, plus we put in a pointer (Figure 5.10). The overhead consists of 5 bytes of the ATM header, 1 byte of regular AAL1 overhead, and then 1 more byte which is the AAL1/SDT pointer. The pointer points to the position of the start of the structure, which then repeats within the AAL1/SDT payload. The pointer is an extra byte of overhead, but we don't put that extra byte in every ATM cell. We use pointers once every eight ATM cells which is equivalent to 1/8 of a byte on average of extra overhead per ATM cell.

Figure 5.11 illustrates the situation for half T1. The pointer points to the start of the 12 bytes of the half T1. Twelve byte structures are inserted into the 46 byte AAL1/SDT payload, so that the last byte is the 10th byte of the 4th structure. The remaining 2 bytes of this structure are transported in the next ATM cell. AAL1/SDT improves the efficiency of voice transport within the ATM network. On the access line, we have an unused portion of bandwidth because we have a fractional T1 service. Half of the T1 access line is unused, whereas inside the ATM network, we are more efficiently utilizing the bandwidth.

We can use partial fill for SDT in order to reduce the delay on SDT communications.

LoS is again detected at the ingress point to the ATM network and in a similar way to the unstructured case, we're able to locate where any T1 faults occurred.

Figure 5.12 gives a comparison between SDT and UDT IWFs. Fault isolation is available from both. We get a gain in networking efficiency in the ATM network with the SDT

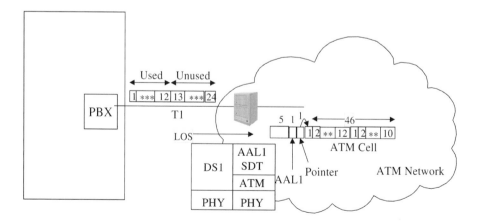

**Figure 5.11**    SDT for 1/2 T1

| | Fault Isolation | Networking Efficiency Gain | AAL1 transmission overhead bytes/ATM Payload | Processing overhead |
|------|------|------|------|------|
| SDT | Yes | Yes | 1.125 | High |
| UDT/ | Yes | No | 1.25 | Medium |

**Figure 5.12** T1 transport SDT/UDT comparison

option, in the sense that a fractional T1 has unused bandwidth on access but not within the AAL1/SDT payload. If AAL1/UDT had been used then any unused time slots on access are also unused within the AAL1/UDT payload. The transmission overhead indicates, for SDT, the number of AAL1 bytes ($=1$) plus one pointer byte every eight cells. For UDT it indicates the number of AAL1 bytes ($=1$) plus approximately 2 T1 overhead bits every ATM cell. The processing overhead is higher for SDT due to the need to calculate the pointer.

This completes our review of the three flavours of AAL1: voice, unstructured, and structured. For just one voice call, we use AAL/voice. For '$N$' voice calls all going to the same destination, e.g. in a private network interconnecting PBXs in different cities, we use SDT. For a full T1 or a full E1 going from, for example, one PBX to another PBX, then we use UDT.

### 5.3.2 AAL2

AAL2 is a means of transporting rt-VBR traffic, e.g. voice traffic with silence detection. AAL2 allows multiple variable bit rate voice calls to share a single ATM virtual circuit. AAL1/SDT and AAL1/UDT perform the same function for CBR voice calls. The standardization of AAL2, [11], was supported strongly by the cellular industry to handle the multiple packetized voice calls arriving over the wireless interface at cellular company base stations. Cellular companies with ATM wired networks need a mechanism for carrying these voice packets over ATM. Each voice packet is typically shorter than an ATM payload so that it would be inefficient to map a single packet into one ATM cell. In order to fill up the payload, the options are:

- wait for more packets from the same voice call, which introduces delay;
- share one ATM cell (and hence one ATM virtual circuit) among several voice calls, which is the option implemented in AAL2.

Each voice call is identified by an AAL2 channel number, so that AAL2 adds a third level of multiplexing to the two levels provided by ATM (Figure 5.13).

Figure 5.14 shows how voice packets are multiplexed into ATM cells. Each packet has a 3 byte AAL2 header attached to it, and the resulting data units are mapped into the ATM payload. An AAL2 data unit may well overlap between two ATM cells, and therefore a

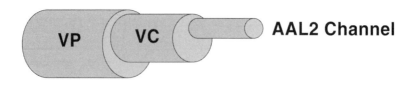

**Figure 5.13**   AAL2 adds another layer of multiplexing

pointer (or offset) is used to indicate where the next AAL2 data unit starts (Figure 5.15). There is some similarity between this pointer concept in AAL1/SDT and AAL2, although the precise format of the pointer is different. As with AAL1, it is not necessary to use the AAL2 pointer in *every* ATM cell, and it is usually inserted once every eight cells.

The format of the 3 byte AAL2 header is shown in Figure 5.16. It contains:

- a Channel Identifier, which identifies the AAL2 channel shown in Figure 5.13. In the cellular company application this corresponds with the individual voice call arriving at the base station.
- a Length Indicator (LI) which indicates the length of the AAL2 data unit. This is required since voice packets using different coding schemes may be of different lengths. The maximum length of AAL2 data units is agreed at call set-up time to be either 48 or 64 bytes.
- a Header Error Control (HEC) to protect the header, primarily the Channel Identifier and the LI. If an error occurs in the Channel Identifier the voice packet is sent to the wrong destination. If an error occurs in the LI, voice packets become garbled until the pointer re-establishes where AAL2 data units start and end.

AAL2 stands for ATM Adaptation Layer 2. However, it can be used over other transport technologies besides ATM. For instance, it can be used to transport packet voice over Frame Relay, HDLC, or IP. AAL2 is a mechanism for taking multiple voice packets from different users and packing them into a common data unit. In fact, AAL2 is simpler to implement over *variable* length technologies like Frame Relay, HDLC, or IP, than it is on *fixed* length ATM cells. The pointer mechanism is required because an AAL2 data unit

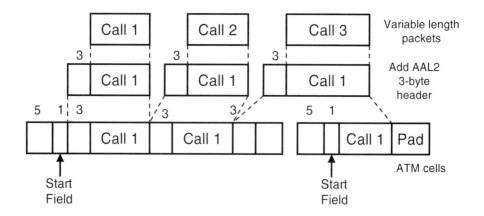

**Figure 5.14**   Variable length voice packets mapped into ATM cells

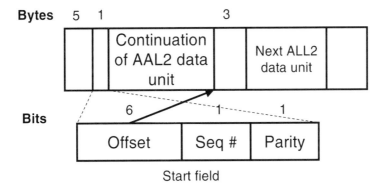

**Figure 5.15**  The start field

may overlap two ATM cells. With a variable length technology like Frame Relay, no overlapping is necessary. The Frame Relay data unit can be adjusted to fit an integer number of AAL2 data units. The pointer is therefore not required, resulting in a simpler implementation.

### 5.3.3  AAL1/SDT Compared to AAL2

Major implementations of voice over ATM focus on the transport of multiple voice calls between an ATM ingress point and an ATM egress point, e.g. between a cellular company base station and a cellular company switching centre, or between customer premises PBXs at different corporate sites. The prime AAL contenders for these applications are AAL1/ SDT and AAL2, and a comparison between them is given in Figure 5.17.

AAL2 achieves more efficient use of bandwidth in the ATM network as a result of allowing silence suppression to be used, which, combined with the statistical multiplexing capabilities of ATM, reduces bandwidth requirements during silence periods. It is important to observe that voice compression can be used both in AAL1 and in AAL2. PBXs

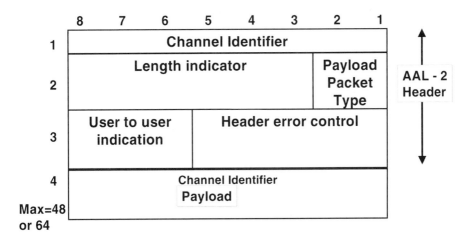

**Figure 5.16**  3 Byte AAL2 header

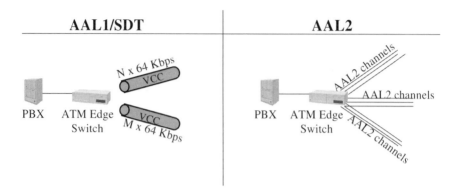

**Figure 5.17**   AAL1/SDT compared to AAL2

compress voice before sending it in fractional T1/E1 time slots. The advantage of AAL2 is that it can handle silence removal which results in a 65% bandwidth saving in addition to the savings due to compression.

The downside to AAL2 is its overhead, which varies depending on the length of the voice packets. As an example, suppose we have 10 byte voice packets, which become 13 bytes after addition of the AAL2 headers. On average 47/13 such data units can be accommodated in a 47 byte AAL2 payload, resulting in the average overhead in an ATM cell being 5 bytes (ATM) + 1 byte (AAL2 start field) + 3 × 47/13 bytes (AAL2 header) = 16.8 bytes. The total number of voice bytes is 10 × 47/13 = 36.1. The percentage overhead is therefore 16.8/53 = 31.8%, compared to an AAL1/SDT over-head of (5 + 1.125)/53 = 11.5%. In practice the AAL2 overhead will be slightly higher than 31.8%, due to the padding used in ATM cells that are not completely full of AAL2 data units.

A comparison between AAL1/SDT and AAL2 therefore needs to take account both of the 65% bandwidth saving due to silence removal and the difference in overhead. Let us take 1 min of voice coded at 8 Kbps using 10 byte G.729 packets. Of this minute 65% is silence. We calculate the total number of bits needed to transport this voice using AAL1/SDT and using AAL2.

- Using AAL1/SDT, we require 60 (seconds) × 8 (Kbps)/(1 − 11.5% (overhead)) = 542 Kbits.
- Using AAL2, we require 35% (non-silence) × 60 (seconds) × 8 (Kbps)/(1 − 31.8% (overhead)) = 246 Kbits. Statistical multiplexing is never perfect, so that, in practice, slightly more than 246 Kbits of capacity in the network will be required, depending on the characteristics of the other traffic on the same links. Also the overhead will in practice be slightly higher due to padding. These two factors are small compared to the difference between 246 Kbits from AAL2 and 542 Kbits from AAL1/SDT.

A comparison of % overhead and the number of Kbits required to transport 1 min of voice is given in Table 5.1 for the major standardized codecs.

The bottom line is that silence removal is more important than differences in overhead, and therefore, given a choice, an implementer would prefer AAL2 to AAL1/SDT from the viewpoint of bandwidth efficiency.

**Table 5.1**  Efficiency of AAL1/SDT and AAL2 for multiple voice calls

| Codec | Kbps | Packet length (bytes) | AAL2 overhead (%) | AAL1 overhead (%) | AAL2 Kbits for 1 min | AAL1 Kbits for 1 min |
|-------|------|------|------|------|------|------|
| G.722 | 64 | 8 | 35.5 | 11.5 | 2084 | 4339 |
| G.723 | 5.3 | 20 | 22.9 | 11.5 | 144 | 359 |
| G.723 | 6.4 | 24 | 21.1 | 11.5 | 170 | 434 |
| G.727 | 40 | 5 | 44.6 | 11.5 | 1515 | 2711 |
| G.728 | 16 | 5 | 44.6 | 11.5 | 606 | 1084 |
| G.729 | 8 | 10 | 31.8 | 11.5 | 246 | 542 |

## 5.3.4  AAL5

There are two distinct ways in which AAL5 is used to transport voice.

1. AAL5 has been selected by the ATM Forum as an option for the transport of voice over ATM at the desktop.
2. Voice traffic which is already formatted into transport packets, e.g. Frame Relay or IP, is sent over an ATM backbone by using AAL5 to segment the Frame Relay or IP packets into ATM cells.

In case 1, the voice is put directly into AAL5 data units. In case 2, the voice is first put into Frame Relay or IP packets and those packets are then fed into AAL5.

AAL5 adds 8 bytes of overhead plus some padding to make the resulting data unit an even multiple of 48 bytes, for transport in ATM cell payloads.

### 5.3.4.1  Case 1: Voice Directly Over AAL5

In order to reduce delay, a maximum of 40 bytes of voice is encapsulated into an AAL5 data unit and then into an ATM cell as shown in Figure 5.18. At the ATM level this results in 25% overhead. AAL5 is therefore used in this way only if bandwidth is low cost, e.g. on a LAN. Low cost bandwidth is the first reason for using AAL5 for voice at the desktop.

The second reason why AAL5 would be used for desktop voice is that it is already there, i.e. at the desktop interface. Any desktop ATM interface must incorporate AAL5 for signalling and for data traffic. Many ATM desktop interfaces were purchased for high bandwidth data applications. Subsequently adding voice for telephony or multimedia communications can be done at a lower cost by using the existing AAL5 interface for voice instead of purchasing a new AAL1 or AAL2 interface.

The disadvantage of AAL5 for voice is that it can only be used end-to-end: from one ATM desktop to another. There is no standard for interworking AAL5 voice into the PSTN for switching to non-ATM destinations. Nor is there a standard for interworking AAL5 voice into AAL1 voice, for the simple reason that the incompatibility between the voice payload lengths (40 bytes for AAL5 and 46/47 bytes for AAL1) results in an interworking delay of over 10 ms, which is undesirable for voice.

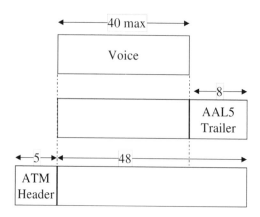

**Figure 5.18**   Voice over AAL5

### 5.3.4.2   Case 2: Voice Indirectly Over AAL5

Many Frame Relay and IP networks have ATM backbones, and AAL5 is used at the edge of the backbone to segment/reassemble the traffic to/from ATM cells. Frame Relay and IP packets that contain voice are not necessarily distinguishable from packets containing data traffic and are therefore sent over ATM using AAL5 by default.

Standardized IWFs for Frame Relay and IP over ATM [4,5] use AAL5, so that even if certain Frame Relay or IP packets were known to contain voice, sending them over AAL1 would result in a non-standard implementation. It is possible to send Frame Relay and IP traffic over AAL2 using the AAL2 trunking standard described below. This is more efficient than AAL5 in situations when the Frame Relay or IP packets are shorter than 40 bytes.

The main reason for sending voice indirectly over AAL5 is therefore that it is simpler than using other AALs. The voice traffic does not need to be distinguished from other traffic and the implementations are standardized making for multi-vendor interoperability.

### 5.3.5   *Voice With Video*

Silent movies are not very popular these days, so video is almost always accompanied by voice. H.261 is a coding standard for videoconferencing which produces $N \times 64$ Kbps of audio and video. It was designed for transport over ISDN. The higher the value of $N$, the better is the video quality and the higher is the cost. A typical value for business video-conferencing is $N = 6$. H.261 is therefore CBR and can be transported over ATM using AAL1.

Two standards are available for packet-based video: H.263 for low quality desktop video and Motion Picture Experts Group (MPEG) for entertainment quality video. MPEG is an asymmetrical video coding standard where you can put in more number-crunching to the encoding than to the decoding. This makes it suited to video retrieval applications, where a lot of processing is done up front to encode the video into a very efficient code. Once this code is stored on a server the video can be retrieved using a less powerful processor. The video quality is dependent on the amount of processing done at the encoding time. For real time applications less processing time is available resulting in a lower video quality. Both H.263 and MPEG can be transported over CBR transmission facilities, but achieve greater efficiency when transported over variable bit rate facilities. The ATM Forum has specified that MPEG video should be transported using AAL5, with AAL1 as a second option.

### 5.3.6  Effect of AAL on QoS

A common misapprehension is that real time traffic cannot use AAL5. In fact, real time traffic should use an appropriate service category, e.g. CBR or rt-VBR, but that service category can be carried over any AAL without affecting its QoS.

Real time traffic can use CBR or rt-VBR service categories. At call establishment time the service category and AAL are specified. PVCs can use any combination of service category and AAL allowed by the network management system. In the case of SVCs, older implementations using UNI 3.1 were restricted in that CBR must use AAL1 and VBR must use AAL5. However, implementations since UNI 4.0 have allowed complete independence between service category and AAL (Figure 5.19).

The service category determines the allowable QoS parameters as described in Section 5.2.3 above. The AAL does not affect which QoS parameters can be specified except in older UNI 3.1 implementations.

The AAL is implemented in silicon and therefore introduces only a very minimal delay. The delay to voice traffic is due to the time taken to assemble the voice packet, which depends on the voice codec being used as described in Chapter 2. The time taken to process the voice packet through the AAL is very small by comparison.

In order to *detect* an error in user data, either AAL5 or AAL1 can be used. AAL1 uses sequence numbers to detect missing cells. AAL5 has got a CRC for bit error detection. In order to *correct* the errors, you have to use AAL1 FEC. So AAL5 will detect bit errors. AAL1 will detect lost cells and the FEC is available, although it is non-standard for voice. For voice, we usually detect errors without correcting them at the ATM layer.

### 5.3.7  Comparison of Adaptation Alternatives

In summary, the selection of AAL for voice depends on the type of voice traffic as shown

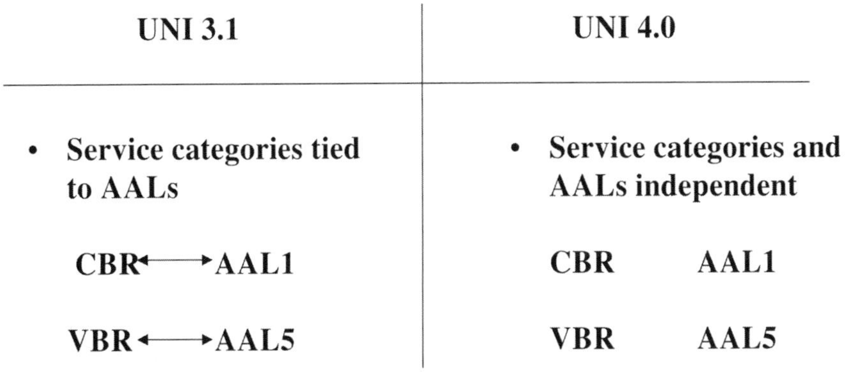

| **UNI 3.1** | **UNI 4.0** |
|---|---|
| • **Service categories tied to AALs** | • **Service categories and AALs independent** |
| **CBR ⟷ AAL1** | **CBR      AAL1** |
| **VBR ⟷ AAL5** | **VBR      AAL5** |

**Figure 5.19**   Adaptation options for CBR and VBR traffic

in Figure 5.20. A single CBR voice call using G.711 at 64 Kbps uses AAL1/voice. Multiple such calls use AAL1/SDT for fractional T1/E1 and AAL1/UDT for full T1/E1. AAL2 is used whenever silence detection is employed in the voice coding, whether for single or multiple calls. AAL5 is an option for use at the desktop and is motivated by being a simple, low cost hardware implementation. AAL5 is also used in standards for interworking with IP and Frame Relay.

## 5.4   ATM Trunking Options for Voice

ATM trunking for voice refers to the use of an ATM network to transport voice between two circuit-based networks or pieces of equipment. It is sometimes also referred to as 'network interworking'.

This does not necessarily imply that the ATM network is in the backbone of the *overall* network. In fact, one trunking option is to use ATM for access between a customer premises and a public network, as shown in the first diagram in Figure 5.21. In this option, CBR voice from a PBX is converted to ATM format at a gateway/IWF on the customer premises. It is then transported over an ATM access network that (a) may be a simple DSL/T1/E1 line or (b) may involve multiple ATM switches or (c) may involve a SONET ring. The gateway/IWF at the other end of the access network converts the voice back to circuit format for long distance transport over the regular PSTN.

The second trunking option shown in Figure 5.21 uses gateways/IWFs at customer premises to interconnect PBXs over ATM virtual circuits that are typically leased from a public carrier.

| | AAL1/ voice | AAL1/ UDT | AAL1/ SDT | AAL2 | AAL5 |
|---|---|---|---|---|---|
| Single voice call 64 Kbps | ✓ | | | | |
| Voice calls with silence detection | | | | ✓ | |
| N x 64 Kbps | | | ✓ | | |
| T1/E1 | | ✓ | | | |
| Desktop voice | | | | | ✓ |
| Voice/ IP Voice/ FR | | | | | ✓ |

**Figure 5.20**   Adaptation alternatives

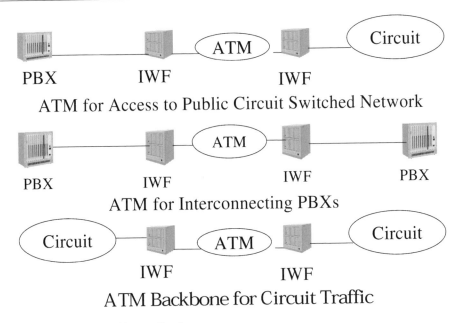

PBX          IWF                IWF
## ATM for Access to Public Circuit Switched Network

PBX          IWF                IWF              PBX
## ATM for Interconnecting PBXs

IWF                IWF
## ATM Backbone for Circuit Traffic

**Figure 5.21**   ATM trunking applications

In each of these two options the benefits of using ATM to transport voice are that (i) it enables voice to be integrated with data on a single access line from the customer premises, thus saving the customer the cost of leasing multiple T1/E1 lines, and (ii) it allows the possibility of saving bandwidth during silence periods when AAL2 is used.

The third option shown in Figure 5.21 is to use ATM in the long haul network backbone to interconnect circuit switched edge networks. For instance, an IXC or other long distance carrier can use ATM to interconnect traffic from local carriers, e.g. ILECs and CLECs. The first implementations of ATM to transport voice were of this type and used AAL1 circuit emulation. They therefore did not benefit from the bandwidth saving of eliminating silence periods which can only be achieved with AAL2. Instead the advantage to the carrier is the simplified network management that comes as a result of having only one backbone network to manage for voice plus all other traffic types. This is a significant attraction compared to the alternative of having separate backbone networks and therefore separate network management systems for circuit switching, Frame Relay, ATM, IP, X.25, etc. Bandwidth saving on the network backbone is of secondary importance since the use of high speed transmission facilities such as OC-48 and OC-192 effectively reduces the $ value of silence periods in individual telephone calls.

There are three ATM Forum approaches to trunking:

- AAL1 circuit emulation [6], which takes CBR circuits and sends them across an ATM network. An extension of this is known as ATM trunking using AAL1 [7];
- AAL1 Dynamic Bandwidth Circuit Emulation Service (DBCES) [8], which again takes CBR circuits and sends them across an ATM network, but allows the capacity of the pipe to vary according to the number of active voice circuits;
- AAL2 trunking [9], which provides a lot of standardized functionality for voice coding and packetization as well as integration with data traffic on AAL2 circuits.

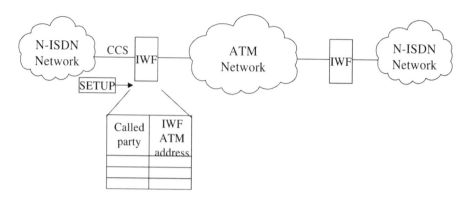

**Figure 5.22**   Connecting IWFs

Each of these approaches is based on assigning ATM addresses to the gateways/IWFs and establishing Virtual Channel Connections (VCCs) between them (Figure 5.22), using either the network management system to set up PVCs or signalling to set up SVCs. A table relates the ultimate destination address of the called party to the ATM address of the egress IWF. An $N \times 64$ Kbps VCC can be used for the voice traffic. There are two options for the signalling traffic which apply respectively to Channel Associated Signalling (CAS) and Common Channel Signalling (CCS).

- CAS involves some of the T1/E1 channels being used for signalling, using 4 bits referred to as the ABCD bits associated with each voice channel. Figure 5.23 shows an example with four voice channels in a multiframe structure. Since the ABCD bits are carried in the T1/E1 channels, there is no need for a separate signalling channel to be set up across the ATM network. Signalling is carried associated with the voice. CAS is commonly used by PBXs.
- CCS is typically used in the public network and involves a separate channel being allocated for signalling. When ATM is used for voice transport a separate virtual circuit is required to carry the signalling. Since signalling is data traffic, it can ride on a nrt-VBR, ABR or GFR virtual circuit, whereas voice needs a CBR or rt-VBR circuit.

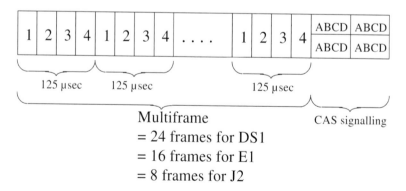

**Figure 5.23**   CAS

### 5.4.1  AAL1 Circuit Emulation and Trunking

The first option for AAL1 circuit emulation is to use UDT, which transports a T1/E1 transparently across an ATM network as described in Section 5.3.1. This is a simple option: a PVC is set up in advance, no signalling is required, and the gateways/IWFs do not need to monitor the individual voice channels as to whether they are actually being used or not. The price we pay for this simplicity is that ATM bandwidth may not be efficiently used at times when the T1/E1 is only partially utilized. The entire T1/E1 is transported at all times.

The second option is to use SDT over PVCs. A fixed group of time slots on a T1/E1 is allocated to each SDT PVC. This is more efficient than UDT, since fractions of the T1/E1 can be transported across the PVC instead of the whole T1/E1. However, the efficiency can still be improved during those times when not all the time slots allocated to a PVC are active, as described in options 3 and 4 below.

The third option is for the ATM network to interpret the non-ATM signalling. Now the ATM network can understand when the non-ATM equipment at either end is requesting a voice channel to be set up or torn down. The gateway/IWF sets up VCCs for the voice according to new call establishment requests. In order to do this it has a mapping between the Called Party Address and the ATM address of the egress gateway/IWF as shown in Figure 5.22. The process can be divided into four steps:

1. A signalling call set-up message is received by the IWF, which determines the ATM address of the egress IWF from its routing tables.
2. If a signalling virtual channel does not already exist and we need to transport CCS, then we set one up. If a voice VCC already exists to that egress point, it is checked for spare capacity to carry the new call. If no spare capacity exists on any existing VCCs, we proceed to step 3.
3. When we get a new call set-up request, we may set up just one 64 Kbps ATM SVC for the new call or we may set up an $N \times 64$ Kbps ATM SVC anticipating other traffic from other users.
4. If CCS is used, we use the signalling virtual channel to pass the set-up message onto the downstream narrowband network. The other thing which we transport is the number of the time slot inside the $N \times 64$ Kbps channel that we are using, i.e. the channel assignment.

In summary, we set up $N \times 64$ Kbps ATM virtual channels according to signalling from the narrowband network. The IWF understands the signalling, and we achieve bandwidth efficiency in the ATM network since the capacity of the VCCs is controlled to match the number of voice calls being transported. There may be some excess capacity planned in the VCC in the ATM network in anticipation of future voice calls needing to be established. If no excess capacity is provided, then each new voice call requires a new SVC to be established. There is therefore a trade-off between (i) providing excess capacity and (ii) requiring a large number of SVCs to be established and managed.

The fourth option is to set up channels according to the time of day. The voice traffic profile typically increases through the day and then declines in the evening. Figure 5.24 shows an example in which we allocate $M$ channels, 24 h a day. At 08:00 h, we add another $N$ channels because people have started to make calls and at 11:00 h we add another $P$ channels. The additional virtual circuits which are set up according to the time of day are

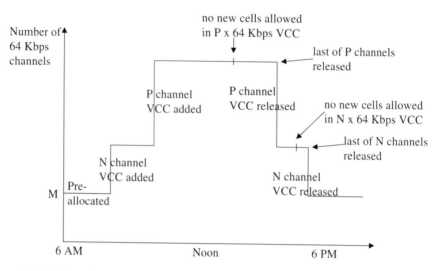

**Figure 5.24**   Time of day VCC assignment

set up by signalling procedures, and therefore signalling messages will request bandwidth from the ATM network just in the same way as the setting up of any other SVC. Then we prevent any more new calls in the $P \times 64$ Kbps channel after a certain time in the afternoon. We wait for the existing callers to finish their calls and once they've all finished using that $P \times 64$ Kbps channel, we can tear down the whole channel. Notice that we have to wait for *all* of the users to finish using it before we can tear it down. The emphasis in the ATM network is on dealing with large amounts of bandwidth instead of dealing with individual channels.

Options 3 and 4 allow bandwidth in the ATM network to be allocated on a call-by-call basis or a time of day basis. When capacity is not used in the ATM network, then we can reallocate it to other traffic. The basic aim is to use ATM bandwidth efficiently so whenever it is not used for voice, it can be used for other traffic. ATM allows the network operator to select the granularity with which he/she wishes to allocate voice bandwidth, i.e. to choose the value of $N$ in the $N \times 64$ Kbps VCCs that are established. A low value of $N$ gives a fine granularity with very efficient bandwidth allocation, but significant processing requirement in establishing multiple SVCs. A large value of $N$ manages bandwidth less efficiently, but requires less SVC establishment. The important point is that the network operator can choose the granularity they wish to use. He/she is freed from the granularity of the non-ATM network where the granularity of E1s, DS1s, E3s, DS3s and SONET/SDH virtual tributaries is fixed and inflexible.

### 5.4.2   Dynamic Bandwidth Circuit Emulation

Dynamic Bandwidth Circuit Emulation (DBCES) allows us to tailor the size of the pipes to the actual traffic requirements from the customers in an even more precise way than we have just discussed in regular circuit emulation. It allows the capacity of the $N \times 64$ Kbps pipes to vary dynamically, i.e. for $N$ to change during the life of the pipe. When a new call

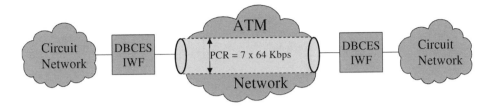

**Figure 5.25**  DBCES Service: call establishment

needs to be set up, the size of an existing VCC can be expanded instead of having to go through the signalling required to establish a new VCC.

We illustrate the operation of DBCES with an example in which there are six voice calls all going to the same ATM egress point plus one channel to carry the signalling messages, making a total of 7 × 64 Kbps.

1. We first establish a 7 × 64 Kbps SVC or PVC (Figure 5.25). The bandwidth in the network is provisioned according to this peak cell rate. We also have policing at the edge of the network at 7 × 64 Kbps to make sure the customer doesn't send too much traffic into the network.

2. The traffic starts off with six active voice channels. The AAL1 header is extended by a bitmask which is one byte, i.e. 8 bits (Figure 5.26). The first six of those bits correspond to the six channels that we're setting up in this connection. The other two bits are unused (unless we have eight voice channels). A one in each of these six bits means that all the six channels are active. As the traffic continues with six active channels, this bitmask is inserted in the cells: not in all the cells, but once every two or four or eight cells. As with the AAL1 pointer it would be an unnecessary use of bandwidth to put it in every cell.

3. Now suppose that one of the telephone users goes on-hook, e.g. channel 3 becomes inactive. The IWF picks up and interprets the narrowband signalling and it knows now that there is no activity on channel 3. It waits until the next ATM cell with a bitmask, and then it puts a zero in for channel 3 (Figure 5.27). It sets the bitmask to indicate which channels are active, and then it starts to generate structures for the active chan-

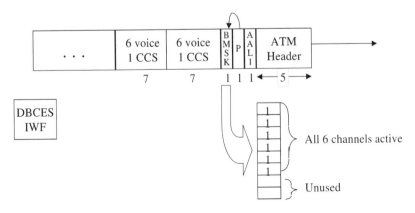

**Figure 5.26**  DBCES: indicating six active voice channels

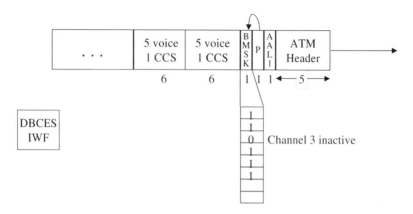

**Figure 5.27**   DBCES: channel 3 becomes inactive

nels. Only five channels are now active, so we just have five channels in the ATM payload plus one channel for the CCS, for a total of six channels.

Regular circuit emulation requires new SVCs to be established to dynamically adjust provisioned ATM bandwidth to the number of active channels. With DBCES, the adjustment is done by changing the bitmask. This is much simpler than the signalling associated with establishment of SVCs. What's new in DBCES is that the allocation of the bandwidth is done dynamically and very simply. We start off with six voice calls, but the IWF notices that one call has finished so that there's only five active voice calls, so we only transport five bytes of voice. The change in the bitmask tells the destination that the upstream IWF is doing this and also informs it which channel is inactive by the information in this bitmask.

The bitmask in the above example is just one byte and so we can only have a maximum of eight channels. If you require more than eight channels, then the bitmask will incorporate more than one byte. You can have a bitmask which extends over several bytes if you've got a lot of channels. Suppose we had 24 possible channels, then the bitmask, when it occurs, will be 3 bytes long.

In order to analyze the benefit of DBCES, we need to carefully assess what other traffic can utilize the bandwidth freed up by DBCES resizing its pipes. Let us first take a public network example. A customer has set up six voice calls. They want to potentially make six voice calls, and are paying for six voice calls. Now one of the calls has gone inactive, so how can the network operator reuse that bandwidth? One option is to use it for any other traffic. Another option is to be careful what it is reused for. The reason for caution is that at any moment the customer may want their sixth channel back. In other words, this is only a temporary reduction in bandwidth requirements. The customer has been provisioned with six channels and cannot be told that part of their bandwidth has been allocated to other traffic. So the network operator can reutilize the bandwidth but can only reutilize it for:

- low priority data traffic like UBR traffic, traffic which does not require any network guarantee either of bandwidth or of QoS;
- other DBCES traffic.

However, the situation is different for an enterprise network. In this case the freed up bandwidth could be reallocated for a temporary increase in throughput for any VBR data traffic until such a time as the user requires it for the CBR, voice application. Although this could technically be done in the public network example above, no additional revenue would be generated, so there is little business reason for doing it.

In summary, the bandwidth is provisioned for the peak traffic. The bandwidth is allocated to the actual traffic, and the question is, 'What can we do with the difference between the provisioned bandwidth and the allocated bandwidth?' We've got $N - M$ channels which can be used for something else. We can use it for UBR traffic or for other DBCES traffic. Some enterprise network operators allow it to be allocated to VBR traffic as well.

### 5.4.3 Summary of AAL1 Trunking Options

In summary, the current focus is on network interworking, where the ATM network is a backbone linking circuit switched access networks. The AAL1 interworking functionality is in three phases. First, set up the multiple PVCs with predetermined bandwidth. Second, set up and tear down SVCs in response to traffic conditions or in response to the time of day. Third, dynamically adjust the size of the virtual circuit by monitoring the channel activity inside the $N \times 64$ Kbps channels and reallocating unused bandwidth to another kind of traffic.

### 5.4.4 AAL2 Trunking

AAL2 trunking adds a significant amount of processing functionality compared to what we have just described for AAL1 trunking, since it deals with compression of G.711 voice into other codes. It handles each individual time slot of the narrowband link separately, whereas AAL1 trunking groups time slots for transport across the ATM backbone. AAL2 trunking becomes involved with all the voice gateway functions described in Chapter 3, including fax and data modem detection, silence removal, dialled digits coding and echo cancellation. It also handles circuit mode data and packet mode data with various degrees of error control.

The IWFs (ATM Forum terminology) or voice gateways (generic terminology) at either side of the ATM network first exchange information about a new voice connection that is being established. Figure 5.28 shows the exchange of information about the codec and packet length. Also specified is the range of sequence numbers to be used for packet loss detection on the voice packets. The error control is provided by the Service-Specific Convergence Sublayer (SSCS), which sits on top of AAL2 in the protocol stack. Each voice packet is given a sequence number within the specified range. If there is silence the sequence numbers are still incremented and that way the destination can determine the length of the silence period because it knows how many sequence numbers were missing. The format is shown in Figure 5.29, including:

- User-to-User Information (UUI), which is used to hold the sequence numbers;
- LI, which gives the length of the payload.

#### 5.4.4.1   SSCS

The SSCS is specified in I.366.1 [12] and I.366.2 [13]. Ref. [12] specifies the segmentation and reassembly of packet traffic and Ref. [13] provides the additional gateway functions,

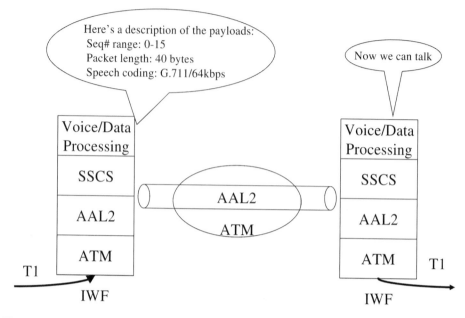

**Figure 5.28**   Specifying the encoding

including fax and data modem detection, silence removal, dialled digits coding, etc. Figure 5.30 shows which types of traffic use which part of the SSCS. CCS can be transported either using I.366.1 or using AAL5.

When I.366.1 is used for packet data, the right part of Figure 5.31 shows the various degrees of error control that it can apply:

- Service-Specific Segmentation and Reassembly (SSSAR) maps long data packets to and from AAL2 data units without providing any error control.
- Service-Specific Transmission Error Detection (SSTED) builds upon the work of SSSAR and adds the capability to *detect* errors. This is suited, for instance, to Frame Relay traffic, where errored data units can be discarded by the network with retransmission provided in end-user equipment, for instance using TCP.
- Service-Specific Assured Data Transfer (SSADT) builds upon the work of SSTED and

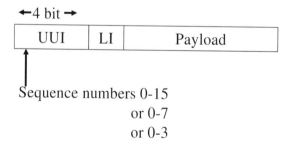

**Figure 5.29**   Sequence numbers assist with synchronization

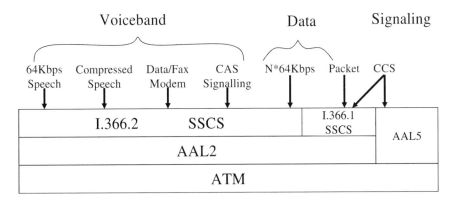

**Figure 5.30**   The IWF can handle voice and data

adds the capability to *correct* errors. This would be suited to signalling traffic for which no other error correction capability is provided at higher layers in the protocol stack.

The left part of Figure 5.31 shows I.366.2 used for speech, fax and circuit mode data, e.g. CBR video conferencing.

The SSCS is isochronous (Figure 5.32) in the sense that it maintains the inter-arrival intervals between packets. The sequence numbers are used to monitor packet arrivals. The inter-arrival times at egress from an ATM backbone are maintained equal to the inter-arrival times at ingress. This is achieved by inserting extra delay when a processing operation completes fast, and by managing a buffer at egress to absorb delay variability introduced by the network, up to a maximum for each specific ATM connection. At times

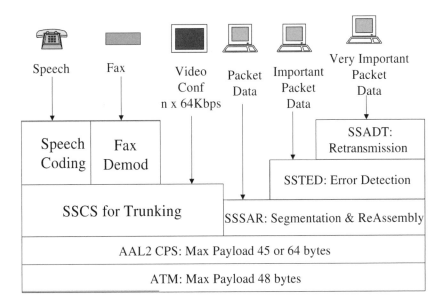

**Figure 5.31**   AAL2 transport of delay-sensitive voice and data

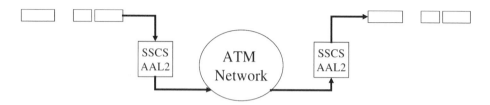

**Figure 5.32**  AAL2 SSCS is isochronous

of congestion, feedback can be provided to the voice codec so that it applies greater compression.

Silence detection can be implemented within a voice codec (e.g. G.723 or G.729) or separately. When the codec itself performs the silence removal, it generates its own Silence Insertion Descriptor (SID). When the silence removal is performed by the AAL2 trunking functionality, the format of the SID is shown in Figure 5.33. Silence is not total silence. There is always background noise, whose amplitude is given by a 7 bit noise level code. Codes 30–127 are defined to represent noise power levels $-30$ to $-127$ dB. During silence, SIDs are sent intermittently, when the amplitude of background noise changes significantly. In that way we maintain continuity in the level of background noise between silence periods and speech.

### 5.4.4.2   Switched and Non-Switched Trunking

AAL2 trunking can be either switched or non-switched [9]. The non-switched case is the simpler one, as illustrated in Figure 5.34, which shows a customer with a PBX in San Diego and another PBX in Atlanta. They want to have four of the time slots on the T1 coming out of the PBX in San Diego always going to Atlanta. The other time slots are to be sent to the PSTN for setting up calls outside the company. The internal company calls between San Diego and Atlanta are on time slots 1 through 4 and get mapped into AAL2 channels in an ATM PVC to Atlanta. The remaining time slots are mapped into AAL2 channels on another PVC to the PSTN. This is called non-switched because it involves using permanent connections across the ATM network. Its not called non-switched because it doesn't go through a switch. It does go through switches. Everything goes through switches; its called non-switched because its not using a SVC.

In non-switched trunking, the traffic in a given time slot (narrowband channel) is always carried in the same AAL2 channel within the same ATM PVC. The signalling messages do not need to be interpreted by the IWF, since the PVCs are already established. Instead they are sent transparently: through AAL2 or AAL5 in the case of CCS. CAS always goes over AAL2 (Figure 5.35). The signalling messages specify the *ultimate* destination of the call. The PVCs are established across the ATM backbone. When the signalling messages reach the other side of the backbone they are passed to equipment that interprets them (the

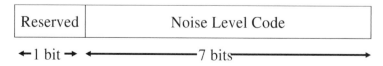

**Figure 5.33**  Generic SID

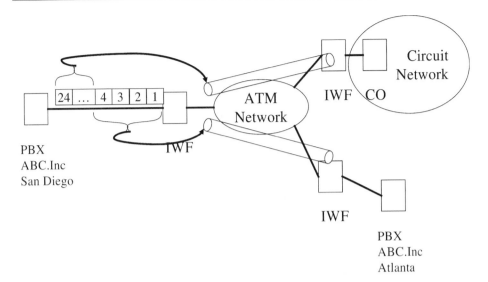

**Figure 5.34** Non-switched AAL2 trunking

customer PBX in Atlanta and the PSTN circuit switch). Neither ingress nor egress gateway interprets them. Non-switched operation is simpler than switched operation and is often implemented first for that reason.

In the switched mode (Figure 5.36) the IWF interprets the signalling messages, figures out where the call has to go and then:

• either routes the call across an existing AAL2 channel within an existing ATM virtual circuit across the ATM backbone;
• or sets up a new AAL2 channel and perhaps also a new ATM virtual circuit in order to get to that destination.

After the call has finished, the AAL2 channel that it occupied may be reallocated to a new call from another time slot on the narrowband network. Thus, there is no permanent relationship between the narrowband time slots and the AAL2 channels, whereas there is such a relationship in the non-switched case.

A similar procedure applies in the opposite direction. When a call arrives from the ATM network, the IWF interprets the signalling messages to figure out the routing across the

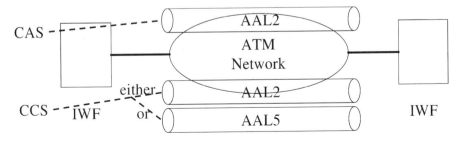

**Figure 5.35** Processing signalling in the IWF non-switched

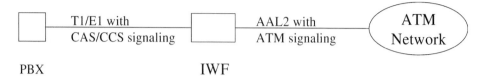

**Figure 5.36**  Switched AAL2 trunking. TDM, Time Division Multiplexed

narrowband network, and then allocates a time slot and a port on the narrowband side of the IWF.

Figure 5.37 illustrates the signalling operations in switched trunking. A signalling message arrives from the narrowband network at the left of Figure 5.37. CAS is converted into CCS for transport across the ATM backbone to the egress IWF:

- either via AAL5 in which case each CCS channel occupies an entire AAL5 channel;
- or via AAL2 in which case it uses AAL2 channel #8, and controls all the voice channels within that ATM virtual circuit.

The ingress IWF interprets the signalling messages and decides on the route across the ATM backbone. If a suitable virtual circuit already exists it is used. If not, the Signalling ATM Adaptation Layer (SAAL) is used to send ATM signalling messages into the ATM backbone to set up an AAL2 virtual circuit (top of Figure 5.37).

An important distinction between the transport of the ATM and CCS messages is that the ATM signalling messages are interpreted by each switch en route across the ATM backbone, whereas the CCS messages are interpreted only when they reach the egress IWF.

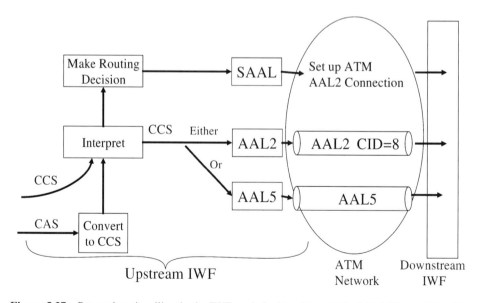

**Figure 5.37**  Processing signalling in the IWF: switched trunking. CID, AAL2 Channel Identifier

### 5.4.4.3   Identifying Idle Channels

We can save a lot of bandwidth on the ATM backbone if the gateway can identify which channels coming in on the T1 into the gateway are idle. This is easy in switched mode because the IWF is reading all the signalling messages and it knows which channels are idle and which channels are active.

In non-switched mode the IWF does not have to read or understand those signalling messages to set up connections, but it could do so in order to monitor the signalling messages so that it knows which channels are idle. This allows it to reduce the bandwidth that is used in the ATM backbone when channels are idle.

The third option is that the IWF is working in non-switched mode and doesn't look at the signalling messages, but it does monitor the channels themselves for idle codes. If it sees a stream of idle codes going down a channel then it knows that channel is not active and it can reduce the amount of bandwidth which is used in the backbone of the network.

### 5.4.4.4   AAL2 Trunking Summary

AAL2 adds a lot of functionality onto what we get with AAL1. The first three options on Figure 5.38 illustrate the functionality of different flavours of AAL1 for trunking. What AAL2 gives us in addition to figuring out which channels are idle is that it detects modems, deals with fax, DTMF and data traffic appropriately, it removes the silence from voice and it allows us to choose a codec which significantly compresses the voice. It is these additional functions which result in a more efficient use of bandwidth in the backbone of the network and, therefore, a better business case for voice over ATM.

It has been the development of AAL2 that has resulted in a good business case for voice over an ATM backbone.

## 5.5   Case Examples

We now analyze how ATM can be used for the same four case examples that we investigated in Chapter 4 for IP transport:

|                  | Idle Channel Suppression | Modem Detection | Silence Removal | Voice Compression |
|------------------|--------------------------|-----------------|-----------------|-------------------|
| CES/UDT          |                          |                 |                 |                   |
| CES/SDT          | *                        |                 |                 |                   |
| DB-CES           | *                        |                 |                 |                   |
| AAL2 Trunking    | *                        | *               | *               | *                 |

**Figure 5.38**   AAL2 trunking compared to alternatives

- desktop applications;
- enterprise networking;
- incumbent public carrier networking;
- competitive public carrier networking.

### 5.5.1 Desktop Applications

There is very little deployment of ATM to the desktop, and even less use of it for voice transport. When it is used there is a choice of AAL5 or AAL1, with AAL5 being the default [10]. AAL5 provides bit error detection, but is heavy on overhead and is therefore suited to calls over a LAN. For calls over a longer distance network, where bandwidth is more costly we use AAL1/voice, which provides cell loss detection and is more bandwidth-efficient. For a voice call using G.711 coding at 64 Kbps, with AAL5 the number of cells per second which we need is 200 plus some OAM cells. With AAL1 we need less cells per second: 171 compared to 200, because we've got less overhead in AAL1 than we have in AAL5. AAL1 can be used for the public network or for the private network, whereas AAL5 is only for the private network.

One option for timing is for the desktop to obtain a clock from the network so that the codec generates code at the speed determined by the network clock. This simplifies the desktop implementation, since it doesn't have to have its own clock, and it does not have to implement adaptive clocking. This reduces the cost of the desktop interface. Alternatively, desktops can implement their own clocks and/or adaptive clocking as options.

A digital technology such as ATM to the desktop involves no two-wire interface and therefore causes no echo at the other end, unless there is an acoustic echo caused by feedback from a speaker to a microphone.

### 5.5.2 Enterprise Networking

Figure 5.39 illustrates the same enterprise voice networking example we used in Chapter 4, prior to the use of ATM. T1 leased lines are used to interconnect PBXs at company sites. Where there is not a direct T1 link, e.g. from Denver to Chicago, traffic is tandemed through an intermediate PBX.

We often have a two-wire interface going from the PBX to an analogue desktop phone, which can cause an echo at the other end. There is therefore a requirement for echo cancellation, and echo cancellers may already be implemented in the PBXs. If not they

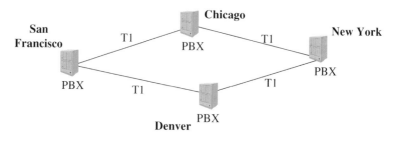

**Figure 5.39**  Enterprise network before ATM

need to be implemented in the ATM equipment at the ingress point to the ATM network in order to control the far end echo.

Figure 5.40 illustrates the use of SDT circuit emulation to utilize the T1 leased lines more efficiently. The T1 interfaces on the PBXs which had in Figure 5.39 been used to interface to the public network are now used to interface to an ATM access device. What we are doing here is to look inside the original T1s to see whether they're full of voice. If they're not full of voice, then we can set up a connection through the ATM network with just the bandwidth that we actually need, and that way we're more efficiently using the bandwidth. Taking into account the AAL1/SDT overhead, we can get a maximum of 21 64 Kbps channels in a T1 using circuit emulation. Suppose that from San Francisco to Denver we've only got, shall we say, 14 phone calls, rather than 21. So we set up an ATM virtual circuit using SDT for just that fraction of the T1 which we actually need to send to Denver. Now we can put some data traffic into the remainder of the T1 bandwidth. Some people would call this free data because we built the network for voice, we found that its not fully utilized for voice and therefore we're putting some data traffic into the remaining band-width. In reality nothing in life is free, and what we pay for more efficient use of the T1 bandwidth is an equipment cost for an ATM switch or ATM access device at each site.

The above analysis is independent of the voice coding used by the PBX. The analysis is based on 64 Kbps channels. The PBX may be compressing two or four voice calls into each 64 Kbps channel. That is an independent function provided by the PBX. The advantage of AAL1/SDT is to efficiently use the T1 bandwidth to transport the 64 Kbps channels independent of how many voice calls they are carrying.

Figure 5.41 illustrates the situation with AAL2. Basically the advantage of AAL2 is that it can handle silence removal. If voice compression has not been performed by the PBX, there is an additional saving in that the ATM access device performs voice coding as well as silence removal. Silence removal reduces the traffic by 60%, so Figure 5.41 shows the sites linked by half T1s instead of full T1s. This is for the case where no additional data traffic is migrated onto the voice network. Alternatively, we could retain the previous network based on full T1s and use 60% of its capacity for data traffic. Again we get 'free data'. Either way we save on the transport cost and that offsets the cost of the ATM equipment needed at each site.

Figure 5.42 illustrates the situation where the customer subscribes to an ATM service

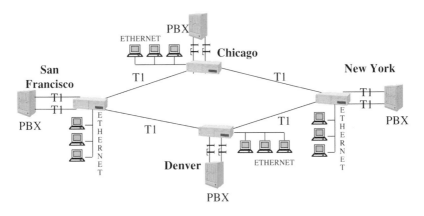

**Figure 5.40**   AAL1/Circuit Emulation Service (CES) plus some data

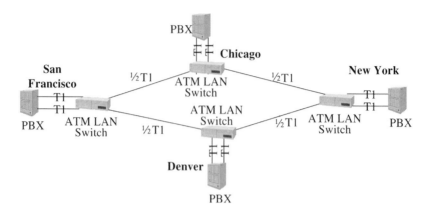

**Figure 5.41**   AAL2 trunking

from a public carrier instead of leasing T1s. The T1 network involved tandem traffic through the customer's ATM access devices, which is removed by subscribing to virtual circuits between each pair of corporate sites. Further efficiency is gained by sizing the virtual circuits to match the actual traffic instead of being restricted to T1/E1s and half T1/E1s.

   In all these options the features on voice calls are provided by the customer's PBX, not by the public network. In the case of smaller customer sites that do not have PBXs, the voice features can be obtained from other PBXs at larger sites in the corporate network or from the public carrier, via class 5 switches or feature servers.

### 5.5.3   Incumbent Public Carrier Networking

The reasons for an incumbent public carrier to use voice over ATM in the backbone are exactly the same as described in Section 4.8.3 for IP. There are seven ATM transport options, which we analyze in order of ease of implementation, starting with the simplest.

1. *AAL1 circuit emulation with PVCs*. The range of interfaces for which circuit emulation is standardized is illustrated in Figure 5.43. We set up AAL1/UDT or AAL1/SDT $N \times$ 64 Kbps pipes across the ATM network between the IWFs and send voice across those

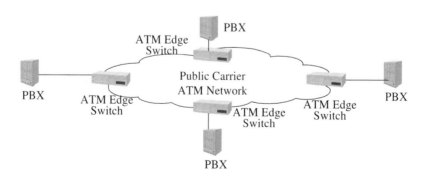

**Figure 5.42**   Virtual private voice network

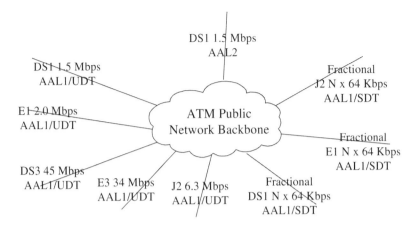

DS1 1.5 Mbps
AAL2

DS1 1.5 Mbps
AAL1/UDT

Fractional
J2 N x 64 Kbps
AAL1/SDT

E1 2.0 Mbps
AAL1/UDT

ATM Public
Network Backbone

Fractional
E1 N x 64 Kbps
AAL1/SDT

DS3 45 Mbps
AAL1/UDT

E3 34 Mbps
AAL1/UDT

J2 6.3 Mbps
AAL1/UDT

Fractional
DS1 N x 64 Kbps
AAL1/SDT

**Figure 5.43**   Incumbent public carrier voice over ATM backbone

pipes. We decide on how big to build the pipes, in other words, what value of $N$ to use, from estimates of the traffic.

2. *AAL1 trunking*. This allows us to set up pipes according to the time of day. Here we monitor the traffic on a typical business day and allocate pipes according to how much traffic we expect at different times of day.

3. *DBCES*. $N \times 64$ Kbps pipes are set up across the ATM network. The IWF monitors any inactivity on the pipes that we have already set up. If it detects channel inactivity, then it informs the destination that the pipe is going to be made narrower. If the channel becomes active once again, the pipe capacity can be increased up to a maximum of the provisioned $N \times 64$ Kbps.

4. *AAL1 circuit emulation with SVCs*. We now start to interpret the signalling from the non-ATM network. This may be SS7 signalling or it may be MEGACO, MGCP, or H.323 signalling. The IWF realizes that the non-ATM user at either end of our ATM network wants to set up a call or they want to tear down a call that they've already set up, and in response to that we set up or tear down SVCs between the IWFs in our network. We may set up individual pipes for individual phone calls or we may set up larger pipes which can accommodate multiple phone calls.

5. *AAL2 trunking, with PVCs*. This is similar to option 1 above except using AAL2 instead of AAL1.

6. *AAL2 trunking, with switched AAL2 channels*. Again we monitor the non-ATM signalling, SS7, MEGACO, MGCP, or H.323, and set up AAL2 channels within existing ATM PVCs using AAL2 signalling.

7. *AAL2 trunking, with SVCs*. This is similar to option 4 above except using AAL2 instead of AAL1.

Interworking is very important for the public carrier network because there is so much technology already in place and any new technology must interwork with the existing technology. Interworking is particularly important between ATM signalling and the SS7 network so as to obtain access to the voice features in the class 5 circuit switches and also to obtain access to advanced intelligent network databases for the routing of toll-free number calls and billing calls to credit cards. This interworking can be done directly between the ATM network and the SS7 network and also via signalling gateways and feature servers in the distributed architecture described in Chapter 3.

### 5.5.4  Competitive Public Carrier

Figure 5.44 illustrates how a new ATM network built in a greenfield situation can be used to carry the customer's voice, Frame Relay and IP traffic. The same seven options are available for voice transport as described in Section 5.5.3. The distinguishing feature of the competitive public carriers situation is that it does not have an installed base of class 5 circuit switches or an SS7 network from which to get voice features. It has the same two options as in the case of IP described in Section 4.8.4: the features can be obtained from the customer's PBX or from a feature server attached to the ATM network.

## 5.6  Conclusion

The first bridge to span the River Thames in London, England was London Bridge completed in 1176 AD, and it was the only such bridge for 550 years. As the city expanded in the 18th century, and cars and trains were invented in the 19th century, more bridges were built to link North and South London. One of the most recent additions is Tower Bridge, 1894, built to show off the ornamentation of Victorian architecture and the power of steam to open and close the roadway. Another is Albert Bridge, 1873, the elegance of its suspension architecture suited to the high-end economic group of 19th century Chelsea which it was designed to serve.

Increasing the number of bridges is only one approach to meeting the demands for increased traffic. Another is to improve and widen the existing bridges. The 12th century wooden structure of London Bridge could certainly not have survived the centuries, and the nursery rhyme about it falling down would probably have come true, had it not been repeatedly rebuilt. The houses that lined its sides, hovering precariously above the murky waters below were removed in these reconstructions and the 1831 version of London Bridge was shipped off as an antique tourist attraction to Lake Havasu in Arizona, to be replaced by the present bridge in 1973.

AAL1 circuit emulation is an old bridge linking the circuit switched world with ATM, and many such bridges were built by Telcos in the mid 1990s. They are simple structures, mapping G.711 voice code into ATM payloads with little processing involved. Increased voice traffic can be served by building additional 'AAL2 trunking' bridges or by upgrading

**Figure 5.44**  New public carrier

some AAL1 bridges to AAL2. AAL2 trunking emphasizes low cost and efficiency, like the sleek style of modern suspension bridges. But the suspension design has been around for over 100 years and there are plenty of older style bridges today that have not been replaced by suspension bridges. Similarly, AAL2 interfaces are being deployed alongside AAL1 interfaces for voice over ATM. AAL2 interfaces were not available on ATM switch ports at the time AAL1 circuit emulation was introduced. Upgrading to AAL2 trunking today would therefore involve throwing out one interface and replacing it by another, the business case for which is not always clear, unless, of course, there is some antique value in an old AAL1 interface.

We can therefore expect a mix of AAL1 circuit emulation and AAL2 trunking for voice over ATM in our networks for some time.

### References

1. ATM Forum User Network Interface Specification.
2. ATM Forum Private Network to Network Interface Specification.
3. ITU-T Recommendation I.363.1 (1996) B-ISDN ATM Adaptation Layer Specification: Type 1 AAL.
4. Frame Relay Forum (i) FRF.5 (1994) O'Leary, D. (Ed.), Frame Relay/ATM PVC Network Interworking Implementation Agreement; (ii) FRF.8 (1995) O'Leary, D. (Ed.), Frame Relay/ATM PVC Service Interworking Implementation Agreement.
5. (i) IETF RFC1483 (1993) Multiprotocol Encapsulation Over ATM Adaptation Layer 5; (ii) RFC1577 (1994) Classical IP and ARP Over ATM.
6. ATM Forum Circuit Emulation Service.
7. ATM Forum ATM Trunking Using AAL1 for Narrow Band Services.
8. ATM Forum (DBCES) Dynamic Bandwith Utilization in 64 KBPS Time Slot Trunking Over ATM – Using CES.
9. ATM Forum ATM Trunking Using AAL2 for Narrow Band Services.
10. ATM Forum Voice and Telephony Over ATM to the Desktop.
11. ITU-T Recommendation I.363.2 (1997) B-ISDN ATM Adaptation Layer Type 2 Specification.
12. ITU-T Recommendation I.366.1 (1998) Segmentation and Reassembly Service Specific Convergence Sublayer for the AAL Type 2.
13. ITU-T Recommendation I.366.2 (1999) AAL Type 2 Service Specific Convergence Sublayer for Trunking.

# 6

# Voice Over Frame Relay

## 6.1 Reference Architectures

Voice over Frame Relay allows voice to enter a Frame Relay network either from a public network channel bank or from a customer premises PBX and is standardized by the Frame Relay Forum [1]. Since Frame Relay has been particularly successful in the access market, a major ingress point for voice into a Frame Relay network is the FRAD in the customer premises (Figure 6.1). Although Frame Relay was originally conceived as a data-only technology, most commercial FRADs today have voice ports, both in the form of T1/E1 interfaces for a PBX and also analogue ports for regular analogue phones over Category 3 wiring.

Most enterprise customers subscribing to Frame Relay public services are migrating traffic from leased line networks (Figure 6.2). The advantages of Frame Relay for voice in this context are:

- With a leased line network there are multiple T1s coming out of the PBX into the network corresponding to the multiple destinations of the voice traffic. As a result of silence removal, less access lines are required from the FRAD to the Frame Relay network, resulting in a reduction in access costs.
- With a leased line network we typically have tandem traffic going through many of the PBXs, e.g. in Figure 6.2 traffic from the PBX on the left to the one on the right needs to traverse one of the other PBXs. With Frame Relay there is no tandem traffic going through any PBX. The reduction in tandem traffic results in:
  - less processing load on the PBX for setting up calls;
  - less switching load on the PBX during calls;
  - less transcoding of voice from one format to another in order to be switched by the PBX, which can result in improved voice quality.

## 6.2 A Frame Relay Primer

Frame Relay is a packet switching technology defined by ANSI (T1.606 [2] and T1.617

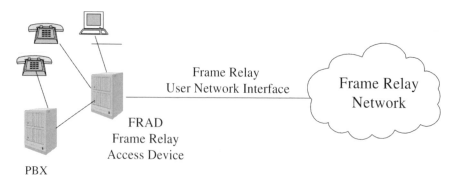

Figure 6.1   Frame Relay architecture

[3]) and International Telecommunication Union – Telecommunication Standardization Sector (ITU-T) (Q.922 [4] and I.122 [5]). The Frame Relay standard specifies a frame, which contains a header, payload and trailer, together with flags that indicate the start and end of each frame. The payload can vary in length up to a maximum determined by the network operator. The ANSI Frame Relay standard specifies that this maximum should be at least 262 octets. Typical North American networks implement maxima between 1024 and 4096 octets. For voice applications short frames of 10–100 octets are used.

### 6.2.1   Virtual Circuits

Virtual circuits are an important concept in Frame Relay since they uniquely identify a connection between source and destination across a Frame Relay network. An end-to-end customer virtual circuit includes:

   (a) the access line from the source customer premises to the ingress frame switch in the public network;

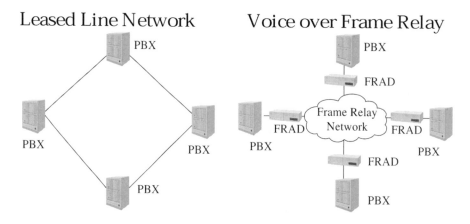

Figure 6.2   Leased lines compared to Frame Relay

(b) the links between one frame switch and another within the Frame Relay network;

(c) the access line from the egress frame switch to the destination customer premises.

In general, many different virtual circuits are sharing each link of type (a), (b) or (c) in a Frame Relay network. Some virtual circuits in general carry voice and others carry data. It would be unusual to mix voice and data on the same virtual circuit, even if they were going to the same destination, since the two traffic types require different QoS. Frames on any given virtual circuit are distinguished from the frames of all other virtual circuits on the same link by a unique DLCI in the frame header. The DLCI used by the source customer is the fundamental means of identifying a connection between a source and destination across a Frame Relay network.

A Frame Relay DLCI remains assigned to a virtual circuit for as long as the subscribers require to continue to communicate between the corresponding source and destination. It requires action on the part of the subscriber to tear down a Frame Relay virtual circuit and hence free up the DLCIs for use by other virtual circuits. For many data applications virtual circuits remain in place indefinitely, but for voice applications, if a virtual circuit is assigned to a single telephone call, it will be torn down when the call ends and the phones go on-hook.

In Frame Relay four DLCIs (1019–1022) are reserved for multicast. A customer wishing to subscribe to multicast service specifies a list of destinations to which an incoming frame should be copied. The frames to be copied are sent from the customer to the network with one of the four multicast DLCIs in their header. Up to four lists can be specified by the customer, one for each of the four multicast DLCIs. At the access switch, the frames with multicast DLCIs are copied into multiple frames and sent to destinations corresponding to the addresses in the lists specified by the customer. If Frame Relay transport is to be used, the incoming multicast DLCI is converted into appropriate DLCIs. Alternatively, other transport technologies can be used inside the carriers network. Frame Relay multicast can be used for voice conference calls, subject to the maximum of four simultaneous conference calls on each access interface.

### 6.2.2  Setting Up Connections

Before any communication can take place over a Frame Relay network a connection must be established between source and destination. The connection corresponds to a virtual circuit from source to destination, and its establishment involves determining the most efficient route and allocating DLCIs on each Frame Relay interface along the route. No physical network resources, in terms of switching or transmission bandwidth, are dedicated to the virtual circuit. Instead, bandwidth is allocated as and when there is traffic to be transported. The network resources permanently dedicated to a virtual circuit are the DLCI values on each interface and entries in the routing tables of the frame switches specifying, for each input DLCI, the output DLCI and output port number.

As in ATM, two options are available for setting up connections in Frame Relay: Permanent and Switched Virtual Circuits. PVCs are available on all public Frame Relay networks, and are set up by means of a service order from a customer and the assignment of DLCIs on each Frame Relay interface between source and destination. Once this procedure is completed, the connection remains in place for several months/years. The second option of SVCs allows connections to be established using signalling procedures following the ITU standard Q.933 [6]. The assignment of DLCIs by means of signalling is suited to connections expected to remain in place for short periods of time.

Initial implementations of voice over Frame Relay used PVCs, because of their widespread availability for data applications. They are suited to the PBX interconnect situation illustrated in Figure 6.2. For dialling to other destinations (off the PBX network) PVCs can still be used between the customer site and a circuit switch, thus avoiding the necessity of deploying Frame Relay SVC software. However, once SVCs are available, bandwidth can be utilized more efficiently by:

- avoiding the use of the circuit switched network;
- using lower bandwidth PVCs between PBXs with SVCs set up to carry overflow traffic.

### 6.2.3   Committed Information Rate

In a circuit-based network, bandwidth needs to be allocated according to the *peak* traffic requirements. Frame Relay, like other packet technologies, allows network resources to be allocated taking into account also the *average* bandwidth of the users data. During the transmission of an individual frame from customer premises to network, peak bandwidth is used, corresponding to the port speed of the FRAD at the customer site. However, there are, in general, gaps between packets on the ingress access line during which the bandwidth requirement is zero, so that an average bandwidth can be calculated and used as a basis for tariffing. In Frame Relay the average bandwidth allowed on an interface is referred to as the Committed Information Rate (CIR).

On a Frame Relay interface multiple virtual circuits are set up: one for each destination to be reached over that interface. A separate CIR is allocated to each virtual circuit, specifying the average bandwidth allowed to the corresponding destination.

For example, a Frame Relay interface may carry three virtual circuits for destinations A, B, C. The interface may be a DS1 access line connected to a 1.5 Mbps FRAD port on a customer premises. The CIRs on the three virtual circuits could be 64, 384 and 768 Kbps, respectively, specifying the average bandwidths allowed to each destination individually.

For voice, the CIR corresponds to the short term average bandwidth generated by the speech codec. Codecs which do not generate voice packets during silence periods can be used to reduce the CIR required and hence the cost of the virtual circuit.

Since multiple CIRs are specified in Frame Relay, a question arises as to limitations on the sum of all CIRs on a single interface. The Frame Relay standard does not specify any such limitation. Ultimately, the maximum traffic is *physically* limited by the port speed of the customer FRAD. The total of all CIRs imposes a *logical* limitation on how CIRs can be selected, rather than a physical limitation on the traffic. Public carriers may specify a limit on total CIR as a percentage of port speed, e.g. limiting the total of all CIRs on an interface to 70 or 150% of the port speed.

### 6.2.4   Access Policing

The specification of average traffic by means of CIR constitutes an agreement between network operator and customer, which needs to be policed by the operator to assess whether the customer is sending excess traffic. Since frames can be sent at the port speed of the customer premises equipment, an excess traffic condition can occur since the CIR is generally at a lower bandwidth.

Frame Relay network operators specify a measurement interval over which bandwidth is to be averaged. Typically averaging takes place over a few seconds or a minute depending on the carrier. For instance, during a measurement interval of 1 s a voice virtual circuit with a CIR of 8 Kbps should not send more than 8 Kbits, which is 1 Kbyte of traffic. The 'allowed burst', $B$, during a measurement interval of $t$ s is:

$$B = t \times \text{CIR}$$

The standard allows carriers to also specify an 'excess burst', $B_e$, in addition to the allowed burst $B$. Carriers choosing not to implement this option are essentially setting $B_e = 0$.

Access policing consists of maintaining a record of the volume of traffic, $T$, on each individual virtual circuit during a given measurement interval. Action is then taken depending on the value of $T$ as follows:

(a) $T \leq B$: frames allowed to pass from the access line on to the network backbone;
(b) $B < T \leq B + B_e$: frames marked 'discard eligible' (see below) and then passed on to the network backbone;
(c) $T > B + B_e$: frames discarded; or
(d) $T > B + B_e$: frames buffered and subsequently discarded if the buffers become full.

In the case where the excess burst, $B_e$, is zero, frames are either accepted or discarded according to the above criteria. Carriers utilizing the option, $B_e \geq 0$, are making use of a single bit in the standard Frame Relay header, known as the Discard Eligibility (DE) bit. Regular frames have DE = 0 and DE is set to 1 for frames meeting condition (b) above. Frames with DE = 1 are transported across the network backbone and are switched by frame switches in a regular manner unless they encounter a frame switch which is experiencing congestion. A congested switch, whose buffers are filling up due to the total volume of traffic it is processing, will discard frames with DE = 1, so as to allocate bandwidth to regular frames. Discarding DE = 1 frames commences when frame buffers reach a specified level.

A simple way of setting these parameters for a voice application is to choose the measurement interval, $t$, to be the length of a 'long' talk-spurt, i.e. a few seconds. On average one can expect a silence period (when the speaker pauses between phrases) during the time $t$, but exceptionally a speaker may be generating voice packets continuously for longer than $t$. Typically, silence periods account for 10% of speech, so that with an 8 Kbps codec, the CIR can be set to $8 \times 90\% = 7.2$ Kbps. With a measurement interval of 5 s, the allowed burst is $8 \times 90\%$ Kbps $\times 5$ s $= 36$ Kbits. If during a measurement interval of 5 s the speaker speaks continuously (with no silence) for more than $36/8 = 4.5$ s, the additional voice packets are marked discard eligible.

DE can be used in a more sophisticated manner with some speech codecs as described in Section 6.4.2.

Use of DE allows a carrier to transport customers' excess traffic as and when network capacity is available after processing regular frames. This may be done only for a specified volume of excess frames by assigning a corresponding value to $B_e$, the excess burst. Alternatively, a carrier may set $B_e$ to a large value so that DE is applied to all excess frames and options (c) and (d) above never come into operation.

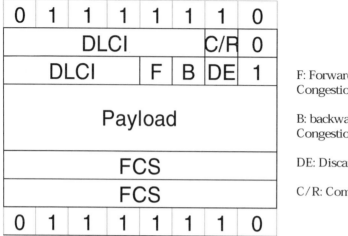

**Figure 6.3**   Frame Relay header format.

Public Frame Relay network operators have several alternative ways of performing access policing:

(i) $B_e = 0$; options (a) and (c);
(ii) $B_e = 0$; options (a) and (d);
(iii) $B_e \geq 0$; options (a) and (b) and ((c) or (d));
(iv) $B_e$ large; options (a) and (b).

### 6.2.5   Data Unit Structure

Figure 6.3 illustrates the Frame Relay data structure, which provides the following functionality:

- *Addressing*. At call set-up time an E.164 address (telephone number) is used to identify the destination. When the connection is established, Frame Relay DLCIs are used to identify it. Technically, the DLCI is not an address. There are not enough DLCIs (in the default header in Figure 6.3) to identify every possible Frame Relay destination in the world. But there are generally enough DLCIs to identify every connection currently established on a given interface. By default there are 10 bits in the DLCI, allowing 1024 connections to be identified. If more than 1024 are required on a given interface, then

**Table 6.1**   Alternative lengths for the DLCI

| Header length (octets) | DLCI length (bits) |
|---|---|
| 2 | 10 |
| 3 | 16 |
| 4 | 23 |

the 2 octet header shown in Figure 6.3 can be extended to 3 octets by setting the Extended Address (EA) in the second octet to zero. The third octet can then be used for an extended DLCI (6 bits) plus a DLCI control indicator plus one more EA bit. If the EA bit in the third octet is set to zero, then a fourth octet is added to the header adding a further 7 bits to the DLCI. The number of bits in the DLCI is given in Table 6.1.

- *Error control.* Error control is performed in Frame Relay by a 16 bit Frame Check Sequence (FCS) which is applied to the frame header and payload taken together. Error control is an important function. It covers both (a) the header, to ensure that the frame is routed to the correct destination and (b) the payload, to ensure that the information that is delivered is correct. The error control is limited to *detection* of errors only. If an error is detected, the corresponding frame is discarded by the network, and recovery is left to layer 4 protocols in the end-user data terminating equipment.
- *Data unit delineation.* Flags are unique bit sequences 01111110, and are used in Frame Relay to indicate the start and end of frames. The remainder of the frame is coded in such a way that it does not contain the flag bit sequence.
- *C/R* is a command/response bit the meaning of which is user-dependent.
- *Congestion control.* A Frame Relay network uses FECN and BECN bits to inform the destination and the source, respectively, that a switch on that virtual circuit is congested. In response, the customer premises equipment may reduce the rate of offered traffic.
- *DE.* In Frame Relay, if a customer exceeds the allowed burst of information within a measurement interval, excess frames may be marked discard eligible. If they subsequently encounter a congested frame switch, they will be discarded to allow regular frames through.

### 6.2.6   Frame Processing

In Frame Relay the variable length frames are switched by a frame switch based upon the incoming DLCI and input port number. We now trace the path of a Frame Relay frame as it is (a) formatted on the customer premises, (b) switched in the network and (c) received by destination customer premises equipment.

### 6.2.6.1   Formatting at Source

A FRAD on the customer premises monitors data packets on the various LANs to which it is connected and voice that is already packetized or in G.711 format in time slots on T1/E1s coming from a PBX. The FRAD packetizes the G.711 and analogue voice and extracts from its input/output tables the appropriate routing for the packets. Packets destined for addresses at a different site are converted to Frame Relay format prior to being sent out of a port on the FRAD connected to an access line into the public network.

If the input/output table specifies that a packet is to be sent over a PVC in a Frame Relay network, then it also specifies the DLCI of the corresponding PVC to be used to reach the required destination. If it is necessary to set up an SVC through a Frame Relay network, then signalling procedures are required, resulting in the assignment of a DLCI value. The frame header is added including the DLCI and, in general, default values for the other fields. The FCS is added, providing error detection on the combined header and payload.

Flags are added at the start and end of the frame and any bit sequence inside the frame which is the same as the flag is modified. The flags now uniquely determine the start and end of the frame, which is transmitted from the FRAD port on to the access line.

If fragmentation is in operation in order to reduce delay to voice packets, it is applied as described in Section 6.5.

In addition to sending frames into the network on the necessary virtual circuits, the source FRAD needs to process information received from the network relating to those virtual circuits. According to the ANSI and ITU standards, congestion control information is sent from the network to the source in a bit in the Frame Relay header used for BECN. The response of the customer FRAD to BECN bits is vendor-dependent. The intention in the standards is that the BECN bit enables the network to inform the user that a certain virtual circuit passes through a congested switch. BECN is a request from the network to the user to reduce the rate of offered traffic on the affected virtual circuit. The FRAD can reduce the traffic flow by one of three methods.

(i) Subsequent frames can be buffered, which requires the FRAD to have sufficient buffer space.

(ii) A message can be sent to the originating equipment (workstation or host computer) requesting it to reduce the traffic flow. This requires the LAN protocol connecting the FRAD to the data terminating equipment to be capable of carrying congestion control information. Frame Relay is an OSI layer 2 protocol, whereas the flow control procedures in data terminating equipment are usually layer 4. The existence of communications between the two is implementation-dependent.

(iii) A higher degree of compression can be applied to speech coding in the FRAD.

For data traffic, many FRADs accept incoming BECN bits, but take no action to reduce the rate of offered traffic. The rationale for this is that there is no need to react to *explicit* congestion information, since the layer 4 flow control software in data terminating equipment is in any case reacting to *implicit* information about network congestion. This can be illustrated by the example of TCP at layer 4 for a data virtual circuit. When a packet is sent into a network, TCP flow control software attaches a sequence number to it. At the destination, a missing sequence number indicates that a packet has been lost. The destination requests a retransmission of the missing packet. When the source TCP software receives a request for retransmission, it retransmits the missing packet. However, several consecutive requests for retransmission are taken to *imply* that the network is congested and is having to discard multiple packets on a given connection. Instead of retransmitting, the layer 4 software reduces the rate of sending traffic to the network. After a certain timeout, the rate of offered traffic is gradually increased. Since network congestion can be deduced *implicitly* by these procedures and reacted to, many users prefer to rely on layer 4 congestion control rather than handling *explicit* congestion notification from a Frame Relay network. The disadvantage in *implicit* congestion control is that it takes place only after congestion has occurred and frames have been discarded, whereas *explicit* congestion notifications can be sent as Frame Relay switch buffers are becoming full, but not after they have overflowed.

For voice traffic, although a FRAD may be able to apply more compression, business customers are reluctant to do this unless they have a contract with their service provider requiring them to do so. They see network congestion as being the service provider's problem, not theirs.

### 6.2.6.2   Processing in the Switch

First the FCS of an incoming frame is recalculated and compared with the value in the trailer of the incoming frame. If the values do not match, the frame is discarded. If the values agree, the frame is accepted as valid and is processed further by reading the DLCI from its header. Access policing is performed to check whether this frame is in excess of the committed or excess bursts allowed on the virtual circuit associated with its DLCI. If so, the frame may be discarded or marked as discard eligible according to the options implemented by the carrier and subscribed to by the user. A look-up procedure is performed in an input/output table taking into account the DLCI value together with the input port number. The result is the priority level of this virtual circuit (vendor-dependent), the output port number and output DLCI value. Once the output port number is assigned, the switch determines whether the associated buffer is congested. If the buffer is completely full, the frame is discarded. If the buffer is at an intermediate level, frames are discarded if they are marked discard eligible. Other buffer levels may also be set to notify the source and/or destination that the network is congested, by setting the BECN and/or FECN bits in the frame header, respectively. At this point the frame header has generally been modified by the switch by (a) changing the DLCI and (b) optionally setting the DE, BECN and FECN bits. It is therefore necessary to recalculate the FCS and insert the new value in the trailer before transmitting the frame on the designated output port. The frame is added to the output buffer and is scheduled for transmission on the port taking into account its priority level.

The priority level is important for voice packets since it can be used to reduce delay in switch buffers.

### 6.2.6.3   Destination Processing

The processing required at the destination customer premises equipment enables the original frame to be extracted from the Frame Relay formatting. In addition, there may be some additional processing to handle congestion control information received from the network.

The first function to be performed is to confirm the error control information contained in the FCS. Once the error control information has been confirmed, the main function is to confirm that the incoming DLCI is valid and to remove the Frame Relay header and trailer so that the original frame can be sent on to the customer premises data or voice equipment. Voice may be decoded in the FRAD and converted to analogue or G.711 format as required.

The other Frame Relay function at the destination FRAD is to respond to congestion control information in the FECN bit of the Frame Relay header. FECN indicates that the path followed by the frame encountered a congested switch. It is a request from the network to the customer to reduce the rate of traffic entering the network on this virtual circuit. It may be used if the FRAD at the destination can control the rate at which the source FRAD is generating frames.

For data traffic using TCP, there is an upper limit on the number of unacknowledged packets that can be outstanding between source and destination at any time. By with-holding acknowledgements, the destination can throttle back the source when the source reaches this upper limit. *Ideally* the receipt of a FECN bit should cause the destination to withhold acknowledgements in this way so that the offered traffic sent by the source on to the corresponding virtual circuit is reduced. After a pre-assigned timeout, and after FECN bits cease to be received, the acknowledgements are released by the destination so that communication recommences. In a *non-ideal* world, this procedure is impractical since Frame Relay is an OSI level 2 protocol, whereas flow control is an OSI level 4 function. The software performing flow control may not be able to communicate with the Frame Relay software, so that, in practice, the flow control software cannot be informed of the receipt of FECN bits from a Frame Relay network.

For voice traffic, the only thing the destination FRAD can do in response to FECN is to signal the source FRAD to apply more compression.

## 6.3  Service Level Agreements

Service Level Agreements (SLAs) are an essential component in the contract between a Frame Relay service provider and a customer. They may incorporate many different items that are negotiated as part of the contract, as described in Chapter 2, some of which correspond with QoS (accuracy and latency measures). It is important to have an unam-biguous method of measuring any item in an SLA, so that the parties to the service contract can agree that it has either been met by the service provider or it has not. The Frame Relay Forum [8] has specified methods of measurement for some of the most common para-meters:

- delay, which is particularly important for voice;
- Frame Delivery Ratio;
- Data Delivery Ratio;
- service availability.

These parameters are measured across a network between a source and destination customer in three possible ways:

1. end-to-end: from customer FRAD to customer FRAD;
2. edge-to-edge egress queue: from ingress point into the public network (first frame switch) to egress point from the public network (last frame switch) excluding the output buffer;
3. edge-to-edge interface: from ingress point into the public network (first frame switch) to egress point from the public network (last frame switch) including the output buffer.

Although the customers' main concern is with 1, it may not be possible for the network operator to have access to the customer FRAD in order to monitor performance at that point. When the customer leases the FRAD from the service provider, 1 is an appropriate measure to use. 2 and 3 are appropriate where the customer controls the FRAD.

For QoS in Frame Relay there is no standard for the vendors as to how they assign priority to the different frames. The only standards are on how to measure the QoS. The standards specify for a service provider how to measure delay, say end-to-end, or how to measure frame loss ratio end-to-end and where it is measured, but they don't specify from the point of view of the equipment vendor how they should achieve a certain end-to-end delay. That's the same for ATM and IP. The ATM and IP standards only specify what types of service quality are defined and how they are to be measured and signalled, not how an equipment vendor should implement service quality. This allows individual vendors to achieve competitive advantage by neat implementations. The implementation is internal to the vendor's equipment and does not therefore need to be standardized.

Delay is measured as a time difference in milliseconds, either for customer frames or for special 'delay measurement' frames which by default are 128 octets long.

Frame Delivery Ratio is measured as:

# of frames delivered/# of frames offered

for three different situations:

* frames within the CIR;
* frames in excess of the CIR;
* total frames.

Data Delivery Ratio is measured in the same way as a ratio of octets:

data delivered/data offered

for the same three situations.

In order to measure service availability, it is important to note that certain service outages are within the control of the network operator and some 'excluded outages' may be defined in the SLA as being beyond their control. During a specified time interval, for any given virtual connection, the service availability parameters are:

* mean time to repair:

    outage time/# of outages

* mean time between outages:

    (time interval − excluded outage time − outage time)/# of outages

* virtual connection availability:

    (time interval − excluded outage time − outage time)/(time interval

    − excluded outage time) × 100%

Any of these four measures of service level can be aggregated, for example:

* over all the virtual connections on a given interface or all the connections for a given customer;
* over a specified measurement time interval, which could be a business day, a 24 h day, a business week, or a 7 day week.

In the case of delay, which is particularly important for voice applications, aggregation can involve averaging delay measurements, e.g. measurements taken every 15 min during the course of a business day. Alternatively, aggregation of delay can involve calculating the percentile of the delay distribution that exceeds a preset level, e.g. the % of measurements (taken every 5 min) that exceed 100 ms on a given virtual connection.

## 6.4 Voice Packetization

The Frame Relay Forum has specified [1] the transport of the output from all the major voice codecs described in Chapter 2. The implementation agreement allows for voice packets from multiple voice calls to be transported in a single Frame Relay frame. We add a sub-channel identifier (CID) to each voice packet, and put that into the Frame Relay frame (Figure 6.4). The Frame Relay frame can therefore end up having a lot of different voice packets with different CIDs in them, plus in principle it could also contain some data packets with associated CIDs. Although it is technically possible to mix voice and data within a single virtual circuit in this way, it is unlikely to be done in practice since each type of traffic requires a different QoS. The end result of using CIDs is an additional level of multiplexing similar to RTP multiplexing discussed in Chapter 3. This is also very similar to AAL2, creating another level of multiplexing. Within an ATM virtual path we get a virtual channel. Then within the virtual channel we get an AAL2 channel. It is similar here. Within the Frame Relay circuit we get a sub-channel. So we can transport each voice channel on a different sub-channel within the same Frame Relay circuit. It has the advantages of:

- reducing the overhead in the Frame Relay network;
- reducing the number of frames that need processing by Frame Relay switches.

### 6.4.1 Sub-Frame Structure

The CID is contained in a header which consists of 1 byte which is required, and then 2 more bytes which are optional (Figure 6.5). If the Extension Indicator (EI) indicator is 1, then we get the second byte. If the LI indicator is 1, then we get the third byte. The main thing that the header does is it provides a space to put in the CID so that we are able to distinguish between the voice on one channel and the voice on another channel, even though both of those channels are in the same Frame Relay frame, and therefore on the same Frame Relay virtual circuit. A single byte header provides 6 bits for the CID and

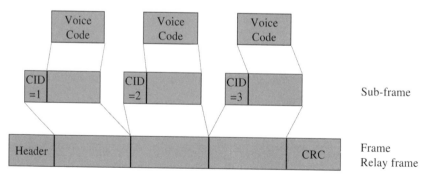

**Figure 6.4** Multiple voice/data sub-channels on a single Frame Relay connection

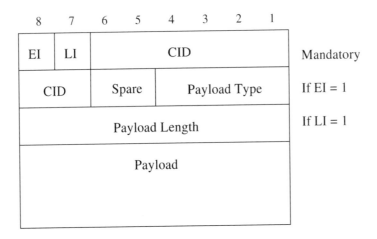

**Figure 6.5**   Sub-frame format

hence has a maximum of 64 sub-channels. A 2 byte header provides 8 bits for the CID and hence has a maximum of 256 sub-channels.

Figure 6.5 also indicates the presence of a Payload Type (PT) indicator, which is set to zero for voice which is regarded as a 'primary payload'. Non-zero values are associated with 'secondary payloads' such as dialled digits (PT = 1) and channel associated signal-ling (PT = 2).

If, as is usually the case, more than one sub-frame is present in a single Frame Relay frame, the 'payload length' is set for each of the sub-frames except the last one. This enables the frame processing software to delineate where one sub-frame finishes and the next one starts.

### 6.4.2   Packing Voice Code into Sub-Frames

The Frame Relay Forum has gone one step further than simply dumping voice code into sub-frames. It has recommended methods of grouping together the most important (core) code so as to distinguish it from the less important (enhancement) code. During conges-

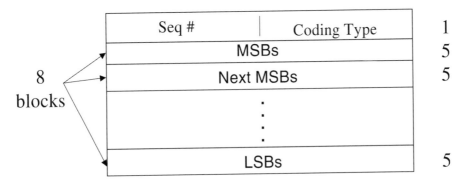

**Figure 6.6**   Voice transfer structure: G.711, G.726, G.727

tion, the enhancement information can be discarded while the core information is transported.

As an example let us take G.711 coding at 64 Kbps over 5 ms, which produces 40 samples of 1 octet each. Instead of simply dumping these 40 octets one after another into the sub-frame payload, we group together the Most Significant Bits (MSBs) of the 40 samples into a group of 5 octets, and put them at the start of the sub-frame (Figure 6.6). The next significant bits go into the next group of 5 octets, etc. and the Least Significant Bits (LSBs) are grouped together at the end. If the network gets congested, then we just cut off the last part of the sub-frame. Effectively we will have applied some voice compression because what we're doing is throwing out the LSBs of the voice code. This is a very simple way of responding to network congestion, and can be used with G.711, G.726 and G.727.

Another way of handling congestion in voice over Frame Relay is that we take the most important voice code, the core information, and we put that in one Frame Relay frame, and we take the less important voice code, the enhancement information, and we put it into another Frame Relay frame (Figure 6.7). The enhancement information gives a better quality of voice, but if we have a congestion condition in the network, then we will throw out the frame with the enhancement information in it and we will keep the core information. So its the core information which will go through the congested switch, and the enhancement information will be thrown out. The way the switch distinguishes between the core information and the enhancement information is from the DE bit in the Frame Relay header. The enhancement information is marked as discard eligible: DE = 1. This mechanism can be used for instance with the G.727 codec which separates out the enhancement information from the core information.

So that gives us two mechanisms in Frame Relay for responding to network congestion by effectively applying increased compression to the voice. One is to throw out the enhancement information from within a sub-frame, and the other is to put the enhancement information into low priority frames and then throw out the entire low priority frame. In each case voice is being used as a cushion which absorbs bursts in the data traffic. Although many public carriers are very concerned about 'toll quality voice' and are reluctant to use voice in this way for *public* telephone service, many *enterprise* network operators recognize that a temporary degradation in speech quality can often be accepted by the human ear, and can be acceptable for intra-corporate telephone calls in return for a lower discard rate (and hence lower retransmission rate) for the data traffic. With voice at about 50% of total traffic on enterprise networks, it provides a very effective cushion for bursty data. As the volume of data traffic increases with a slower rate of growth in voice, the proportion of voice traffic becomes smaller and its cushioning effect becomes more limited.

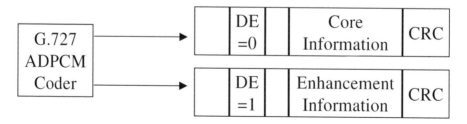

**Figure 6.7**   Mapping enhancement information to low priority frames

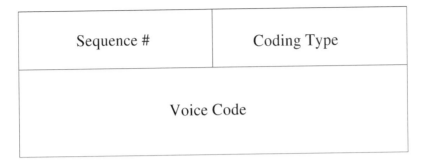

| Sequence # | Coding Type |
|:----------:|:-----------:|
| **Voice Code** | |

**Figure 6.8**   Voice transfer structure: G.723.1, G.728, G.729

G.723, G.728 and G.729 produce voice code that is not divided into core and enhancement information, so that the above techniques cannot be applied. However, these codecs produce code which is already highly compressed, so that it is unrealistic to expect additional compression in any case. The format for transporting it over Frame Relay is shown in Figure 6.8. For instance, G.729 produces 10 octets for each 10 ms voice packet. Between one and six G.729 voice packets from a single phone call are packed into a single sub-frame. G.728 produces 5 octets for each 2.5 ms voice packet and these packets are handled in pairs, i.e. 10 octets for each 5 ms of voice. Between one and 12 such pairs are packed into a single sub-frame. G.723.1 produces either 24 octets (6.3 Kbps) or 20 octets (5.3 Kbps) every 30 ms. A single packet of this type is transported in a single sub-frame. These alternatives are summarized in Table 6.2.

When the voice codec detects silence and stops generating voice packets, a Silence Insertion Descriptor (SID) (Figure 6.9) can be transported. It performs two functions.

1. It is used to describe the amplitude of the background noise. Playing out total silence at the destination is unacceptable since the listener experiences a discontinuity from the background noise that was present during speech activity. It is therefore important to play out background noise of the same amplitude during 'silences'.
2. It is used to indicate the length of the silence period. This is accomplished by using sequence numbers, which are incremented every 5 ms modulo 16. The sequence numbers are also transported in the sub-frames containing voice code (Figures 6.6 and 6.8). The source continues to generate sequence numbers during silence periods and to transport some of them in SID packets so that the destination can deduce the length of the silence period with a minimal amount of information being transported during silences.

**Table 6.2**   Voice code packetization

|              | Packet size (octets) | Packet duration (ms) | Packing factor |
|--------------|:--------------------:|:--------------------:|:--------------:|
| G.723.1      | 24                   | 30                   | 1              |
| G.723.1      | 20                   | 30                   | 1              |
| G.728        | 10                   | 5                    | 1-12           |
| G.729 Annex A | 10                  | 10                   | 1-6            |

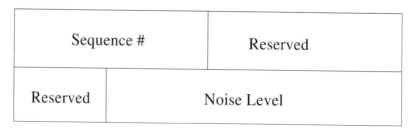

**Figure 6.9**   Silence Insertion Descriptor (SID)

The sequence numbers in speech packets can also be used to detect lost sub-frames so that the destination can use a concealment technique to compensate for the loss, as described in Section 2.2.3.

A complete picture of the packetization of voice over Frame Relay is shown in Figure 6.10 which incorporates:

- the Frame Relay header and trailer;
- a single sub-frame header (in general many could be used);
- the voice code header.

In summary, voice over Frame Relay can transport speech, dialled digits, signalling, fax and voice band data. It is primarily used in the access network. Voice is transported in sub-frames inside the Frame Relay frame with different voice calls on different sub-channels, and when the network gets congested, the enhancement information in the voice code can be discarded. Short data frames should be used on low bandwidth interfaces, which can be achieved using Frame Relay fragmentation, which is discussed next.

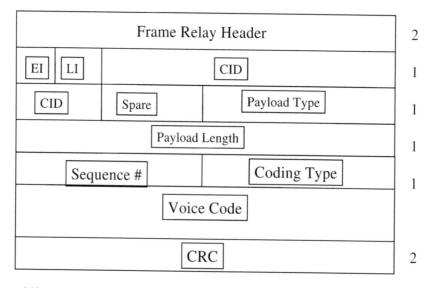

**Figure 6.10**   The whole frame

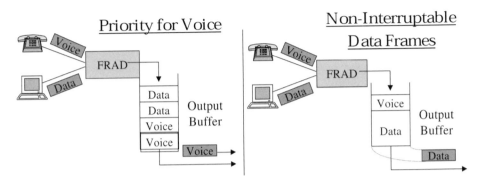

**Figure 6.11**   Reducing delay for voice over Frame Relay

## 6.5   Frame Fragmentation

### 6.5.1   Benefits of Fragmentation

Voice over Frame Relay requires a tight end-to-end delay to control the echo and also to maintain the interactivity of human conversation. Let us suppose that our Frame Relay network is transporting both voice and data. The voice and data frames drop into the output buffer of the FRAD before going off into the Frame Relay network. In order to give a low delay to the voice, we can give the voice traffic a higher priority than the data traffic, so that the voice traffic gets sent out on the transmission line first. That's fine, except for the case where we have a data frame in our buffer which is *already in the process of being transmitted* onto the output port (Figure 6.11). If this data frame cannot be interrupted, it can cause a significant delay to the voice. Suppose the data frame is 4000 bytes long; the delay to the voice frame depends on the speed of the transmission facility. If it is an OC3, its not going to take long for the data frame to get out the door, and we don't incur much delay on our voice traffic. If it is a T1, and we assume that on average the voice frame needs transmitting half way through the data frame, then it will have to wait for 2000 bytes at 1.5 Mbps, i.e. 10 ms, which is tolerable, especially in the presence of echo cancellers. The problem comes if the customer access line is slower than T1. If its a 64 Kbps channel, then the voice delay is about 250 ms, which is intolerable.

There are three possible approaches to the issue of data frames causing delay to voice frames:

1. use high bandwidth interfaces, at least half T1;
2. retune the maximum transfer unit length on all data applications riding over Frame Relay so as to reduce the impact of data traffic on voice delay;
3. implement Frame Relay fragmentation [7].

Option 1 is available for many Frame Relay sites with sufficient traffic volumes to justify high bandwidth access lines. However, these large customers often need to bring smaller branch offices with lower bandwidth access lines into the corporate network.

Option 2 involves a lot of work for the data communications staff of a company since many applications and higher layer protocols are often transported over Frame Relay.

Hence, option 3 is a necessity for the migration of voice traffic onto Frame Relay

## Without Fragmentation          With Fragmentation

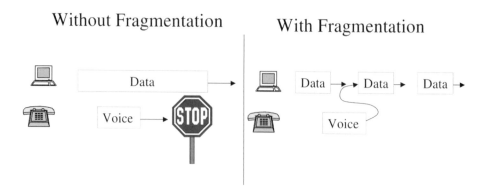

**Figure 6.12**   Frame Relay fragmentation

networks that had previously been transporting only data. This is the option which we now describe in detail.

With fragmentation the long data packet is chopped up into smaller fragments so that the voice packet can get in between two of those fragments (Figure 6.12). The Frame Relay Forum has standardized the method of fragmentation [7], which is particularly useful in situations where we might have a significant delay to the voice packet.

### 6.5.2   Fragmentation Formats

Figure 6.13 illustrates three implementations of Frame Relay fragmentation: one is on the UNI, one is on the Network-to-Network Interface (NNI), e.g. between two Frame Relay network operators, and the other is end-to-end. The advantage of fragmentation is that it solves the problem of delay on low speed interfaces. An end-to-end connection will in general traverse both high and low speed interfaces. The only reason, therefore, for

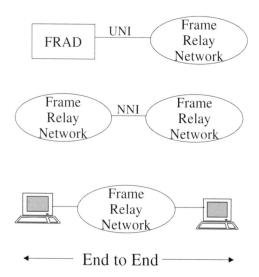

**Figure 6.13**   Frame fragmentation: three implementations

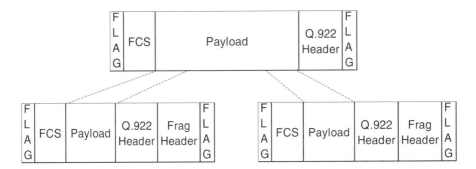

**Figure 6.14**  UNI/NNI fragmentation procedure

implementing fragmentation end-to-end is if there are some low speed interfaces in the end-to-end path that do not implement fragmentation. Our first preference would be to implement fragmentation only on those interfaces that need it: the low speed ones. If, for some reason, fragmentation is not available there, it needs to be implemented end-to-end.

On a low speed interface, fragmentation must be applied to all virtual circuits. In an end-to-end situation, fragmentation needs to be applied only to certain virtual circuits, namely those that traverse low speed interfaces during their journey through the network. This gives rise to different implementations for the fragmentation procedure.

- on the UNI and NNI: where all frames are fragmented, the fragmentation header comes before the Frame Relay header since the receiving equipment is expecting fragments;
- end-to-end: where some frames are fragmented and some are not, the Frame Relay header comes before the fragmentation header since the receiving equipment is expecting frames, and it does not know in advance which frames are going to be fragmented.

Figure 6.14 shows the first case, where all frames are fragmented. Basically, the fragmentation procedure takes the Frame Relay frame which has got flags at either end, a header, a payload and a trailer, and it chops that up into fragments with a fragmentation header sitting in front of the regular Q.922 Frame Relay header. This fragmentation procedure is suitable in a situation where all of the frames on a given interface are being fragmented, which corresponds to the UNI and the NNI interfaces.

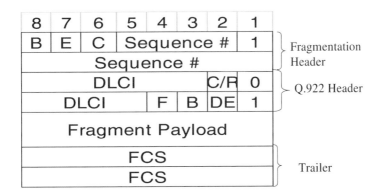

**Figure 6.15**  UNI/NNI fragmentation header format

The UNI/NNI fragmentation header (Figure 6.15) consists of:

- a 12 bit sequence number, which is used to track missing fragments. Separate sequence numbers are used for each virtual circuit on the interface;
- indicators called B and E which are the beginning and ending indicators. If we take a long Frame Relay frame and fragment it into a number of different fragments, the B indicator is set to 1 for the beginning fragment and the E indicator is set to 1 for the ending fragment. That way the destination equipment knows how to reassemble those fragments into the original frame;
- an indicator C which is reserved for future use;
- the LSB of the first byte is set to 1. This distinguishes the fragmentation header from the Frame Relay header which has a zero in that position. This bit can be used by the equipment at one end of the UNI/NNI to detect misconfigured equipment at the other end of the UNI/NNI. The equipment at both ends must either be using fragmentation on all frames or on no frames.

On the other hand when we're implementing fragmentation end-to-end, since only certain frames are going to be fragmented, we put the Frame Relay header first and then a fragmentation header later on in the payload of the Frame Relay frame (Figure 6.16). This follows a coding system defined by the Frame Relay Forum which allows a Network Layer Protocol Identifier (NLPID) to be assigned to the traffic in the Frame Relay payload, FRF3.1 [9]. The NLPID 10110001 has been assigned for Frame Relay fragmentation so whenever the destination equipment sees that, it knows that it should expect a fragmentation header to follow. This way we can have some virtual circuits with fragmentation and some without. The fragmentation header follows the same format as in Figure 6.16 except that the LSB of the first byte is zero, since the issue of detecting a misconfigured interface does not arise. The individual switches on the Frame Relay end-to-end virtual circuit just read the Q.922 Frame Relay header and switch the frame accordingly. They do not look at the FRF3.1 header or the fragmentation header. Only the destination equipment at the end of the virtual circuit looks inside the Q.922 payload and interprets the other headers, reassembling the fragments into frames.

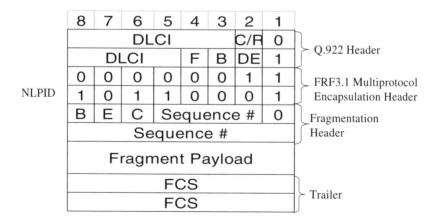

**Figure 6.16**   End-to-end fragmentation header format

### 6.5.3   Overhead/Delay Trade-Off

Fragmentation adds in two ways to transmission overhead:

(a) first, there are additional header bytes:

- two in the UNI/NNI case;
- four in the end-to-end case.

(b) second, since the fragments are smaller than the original frames the headers occur more frequently. A Frame Relay frame with a 1500 byte payload contains two flags, a 2 byte header and a 2 byte trailer, giving a total transmission overhead of 6/1506 = 0.4%. The same frame, segmented into 50 byte payloads, results in 30 fragments, each with 8 bytes of overhead on the UNI/NNI and 10 bytes of overhead in the end-to-end implementation. The percentage overhead is 8/58 = 14% and 10/60 = 17%, respectively.

Since voice packets are usually short, fragmentation adds to their transmission overhead, only due to factor (a) above. The major effect is due to (b) on data frames.

The trade-off between overhead on data frames and possible delay to voice frames is shown in Figure 6.17 for the UNI/NNI. The possible delay is due to the emission time of a data frame, and will occur if a voice frame arrives immediately after a data frame starts to be transmitted.

On a low speed access line, for instance a 64 Kbps access line into a Frame Relay network, if a voice packet encounters a 1500 byte data packet just ahead of it, the voice packet could get delayed by 188 ms, which is very significant for just one hop in its journey through the whole network. On the other hand if that 1500 byte packet was fragmented into 50 bytes frames then the delay for each fragment would only be 7.2 ms which is much more acceptable. So, particularly on low speed interfaces, if we want to mix voice and data traffic together, it makes sense to fragment the data frames into smaller fragments. The price we pay for fragmentation is an increase in transmission overhead, since each fragment requires its own header. In the above example, reducing the delay from 188 to 7.2 ms is accompanied by an increase in the transmission overhead from 0.4 to 14%. The overhead is independent of the speed of the interface, but the delay is higher for low speed interfaces.

## 6.6   Case Example

In the public network Frame Relay is generally regarded as an access technology. Therefore, the organization most likely to put voice over Frame Relay is the customer as

|  | 64 Kbps | 384 Kbps | 1.5Mbps |
|---|---|---|---|
| 1500 byte payload non-fragmented |  |  |  |
|     Emission time (msecs) | 188 | 31 | 7.8 |
|     % overhead | 0.4 | 0.4 | 0.4 |
| 50 byte fragement payload |  |  |  |
|     Emission time (msecs) | 7.2 | 1.2 | 0.3 |
|     % overhead | 14 | 14 | 14 |

**Figure 6.17**   Fragmentation: overhead/delay trade-off

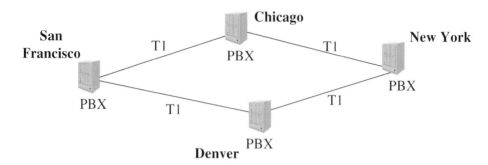

**Figure 6.18**   Enterprise network based on leased lines

opposed to the carrier. Thus, the only case example we will consider for voice over Frame Relay is the enterprise network.

### 6.6.1   Enterprise Networking

The issues relating to sending packet voice over an enterprise Frame Relay network are very similar to those involved in sending voice over an ATM network using AAL2, as described in Section 5.5.2. Figure 6.18 gives the reference configuration of sending voice between PBXs using T1s.

Figure 6.19 shows the use of voice over Frame Relay, where the voice is packetized in FRADs at the customer premises. A standard voice coding can be selected such as G.723 or G.729, which incorporate silence removal, or G.727 can be used putting the enhancement information into DE = 1 frames. The PBX sits behind the FRAD. We set up PVCs between the customer's site for the voice traffic just the same way that we do for the data traffic. The customer needs to select a carrier with a good SLA that will give short delay to the voice traffic. Also if there is a significant amount of off-net traffic, the customer can benefit from a carrier that can offer gateways that take the customer's voice from the Frame Relay network at a convenient point and interwork it onto the PSTN using G.711 coding.

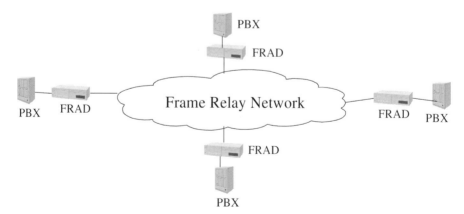

**Figure 6.19**   Enterprise network using voice over Frame Relay

If the customer needs to integrate the voice traffic with data traffic, which is highly likely, since most Frame Relay networks were originally built for data, then the customer's FRAD needs to incorporate Frame Relay fragmentation on low speed UNIs between the customer and the carrier.

The carrier perspective on this is that they need to implement QoS so that the voice virtual circuits get a better end-to-end delay than the data circuits and also the carrier can use voice as a cushion to alleviate data congestion by dropping out the less significant bits of the voice code.

## 6.7   Conclusion

The water reaching the Aswan Dam in Egypt originates partly from the East African Plateau via the White Nile and partly from the Ethiopian mountains. Although the rainfall in East Africa is seasonal, by the time the White Nile reaches Khartoum in Sudan, the flow has been evened out by the buffering effect of the East African lakes and the Southern Sudanese swamps through which it has passed. By contrast, the spring rains in Ethiopia cause an immediate torrent in the Blue Nile, causing it to rise 20 feet where it joins the White Nile in Khartoum. Such is the force of the Nile flood at Khartoum that some of the White Nile water is held back in a temporary lake. The seasonal variations in the volume of water from the Blue Nile control the mix of water from each river that proceeds downstream from Khartoum to Aswan. Lake Nasser at Aswan has a capacity of 40 cubic miles which is used to control the flow further downstream to Cairo.

The traffic arriving at the edge switch of a Frame Relay network originates partly from the customer's desktop data equipment via a LAN and partly from their desktop phones, via a PBX. Although the data traffic from the end users is bursty, by the time it reaches the FRAD it has to some extent been evened out by the LAN switch buffers through which it has passed. By contrast, the voice traffic is not delayed by the PBX and rushes immediately to the FRAD. The voice traffic zaps through the FRAD, holding back some of the data traffic in buffers. The millisecond by millisecond variation in volume of voice traffic dictates the mix of voice and data traffic that travels down the access line from the FRAD to the service provider's edge switch. Here more buffering takes place to control the flow into the network backbone.

## References

1. Frame Relay Forum Voice Over Frame Relay Implementation Agreement FRF.11.
2. ANSI T1.606 (1990) Architectural Framework and Service Description for Frame Relay Bearer Service.
3. ANSI T1.617 (1991) Signalling Specification for Frame Relay Bearer Service for DSS1.
4. ITU-T Q.922 (1992) ISDN Data Link Layer Specification for Frame Mode Bearer Services.
5. ITU-T I.122 (1993) Framework for Frame Mode Bearer Services.

6. ITU-T Q.933 (1995) Signalling Specifications for Frame Mode Switched and Permanent Virtual Connection Control and Status Monitoring.
7. Frame Relay Forum FRF.12 Frame Relay Fragmentation Implementation Agreement.
8. Frame Relay Forum FRF.13 Service Level Definitions Implementation Agreement.
9. Frame Relay Forum FRF.3.1 Multiprotocol Encapsulation Implementation Agreement.

# 7

# Comparison Among Alternative Transport Technologies

The three main packet transport technologies for voice each have their own advantages and disadvantages, which we delineate in the current chapter. Our aim here is to present a *very concise summary* of many points that are dealt with in more detail in Chapters 1–6, organized according to strengths and weaknesses of each technology. This chapter can be regarded as an Executive Summary of the earlier part of this book, and readers are referred to those chapters for the details and also for the expansion of the acronyms used here.

## 7.1  IP

### 7.1.1  Strengths

*Widespread deployment*. IP is everywhere, from the desktop, through the LAN and access networks, to the network backbone. The proliferation of the public Internet has made it an essential transport technology for all traffic types including voice.

*Desktop implementation*. In particular, desktop deployment of IP results in direct customer access to software to control calls, add features to calls and integrate voice calls with other traffic for multimedia conferencing.

*Web integration*. The widespread use of the Web on the public Internet and on corporate intranets facilitates the integration of voice over IP with the Web for call set-up, access to features and Web-based call centres.

*IP phones*. The availability of wired and wireless phones that send voice over IP from the handset implies the importance of IP transport over the whole network. The implementation of IP PBXs contributes to the same end result.

*Header compression*. Header compression is very important for voice in reducing RTP/UDP/IP transmission overhead. IP was designed to transport long data packets. Header compression is essential to the efficient transport of short voice packets.

*MEGACO*. MEGACO has emerged out of a number of signalling proposals at the IETF, and is efficient in the connectionless world of IP.

*SIP*. SIP is an IP protocol that provides 'find me' and 'call forwarding' functions that are increasingly important in the mobile environment that many users are accustomed to.

### 7.1.2   Weaknesses

*QoS*. RSVP and DiffServ are available for providing QoS in an IP environment and each has its own limitations compared to the comprehensive QoS requirements described in Chapter 2. RSVP allows the user to quantitatively specify delay requirements, but not delay variability or accuracy. DiffServ allows the user to request a priority level with respect to accuracy and delay, but it does not deal with delay variability, nor does it allow the user to request quantitative values for accuracy and delay.

*H.323*. H.323, a long-standing signalling protocol for real time traffic over IP, is implemented in many IP call control servers, but is being supplanted by MEGACO.

## 7.2   ATM

### 7.2.1   Strengths

*QoS*. One of the main strengths of ATM is its QoS capabilities, which are standardized and well proven in multi-vendor commercial implementations. ATM allows the user to request *quantitative* values for the three major elements of QoS: accuracy, delay and delay variability.

*Installed base*. All major carriers including incumbents, competitive carriers and Internet service providers have a significant installed base of ATM equipment which was originally deployed in the core of the network. As competing technologies such as IP and Dense Wavelength Division Multiplexing (DWDM) have been introduced, network operators have been reluctant to give up their investment in ATM equipment, but instead have made it interwork with the competing products. As DWDM was introduced in the core of the network, ATM switches were moved from the core to the edge. As high throughput IP switches became available with the same capacity as ATM switches, the focus moved from a single technology to equipment that could interwork with both IP and ATM, for instance using MultiProtocol Label Switching (MPLS). Voice needs to be transported over ATM, therefore, simply because there is a large installed base of ATM equipment.

*Widespread deployment*. Associated with the previous point is the regions of the network in which ATM is deployed, including widespread deployment in both the network backbone and also the access network.

*Network management*. ATM has comprehensive network management capabilities that are well proven in the industry. Network management capabilities for performance monitoring and fault isolation are available in standardized interworking functions. The format of data in Management Information Bases is standardized allowing access by all network management systems, in particular using standard SNMP. Well developed and standardized network management is another reason for expecting ATM to be around for some time.

*Connection orientation*. ATM is connection oriented, which suits the operating environment of the incumbent and competing carriers, since it provides pipes that can be managed. Pipes between customer sites and between network gateways can be sold, policed, tested, and billed.

*Adaptation options.* ATM provides a range of adaptation options from which the user can choose. AAL1, AAL2, and AAL5 can each transport voice in a different way, thus allowing the user to tailor the ATM service to their own needs.

*AAL2.* AAL2 in particular offers many options for transport of speech, fax, modem data, DTMF, and CBR and VBR data.

*Signalling.* ATM offers two signalling options:

- the complex UNI signalling that offers comprehensive call set-up with associated QoS;
- the simple AAL2 signalling that can be used to set up an AAL2 channel within an existing virtual circuit.

Combining the two brings the advantages of both worlds, with UNI signalling used once up front to give QoS, and AAL2 signalling used for individual phone calls.

### 7.2.2   Weaknesses

*High overhead.* The 'cell tax' of a high percentage transmission overhead is often cited as a disadvantage of ATM. This cell tax becomes even higher when adaptation overhead is added, and in particular AAL2 is heavy on overhead. However, AAL2 allows silence periods to be removed from speech offsetting some of the effects of high overhead. Section 7.4 provides a comparison among IP, ATM and Frame Relay from the viewpoint of transmission overhead.

*MEGACO modifications.* When ATM is used as the transport technology for MEGACO, additional signalling is required to set up ATM SVCs.

## 7.3   Frame Relay

### 7.3.1   Strengths

*Installed base.* Throughout the 1990s the market for Frame Relay grew rapidly resulting in a significant installed base of customers subscribed to the service, with customer premises FRADs and Frame Relay interfaces on edge network equipment.

*Bandwidth efficiency.* Frame Relay itself has low transmission overhead, and when used for voice, minimal additional overhead is required.

*Voice encapsulation.* The standards for encapsulating voice into Frame Relay frames has been done in such a way that the less important parts of the voice code can be dropped in the case of network congestion.

*Voice as a cushion.* During periods of congestion in a Frame Relay network, voice can be used as a cushion for bursts of data traffic, with the least significant parts of the voice code being dropped instead of dropping a data packet. This is suited to the fact that most voice over Frame Relay implementations are in enterprise networks. The data communications manager can make a trade-off between voice quality on *internal* calls and accuracy on data transport. On the public network, with customers paying for 'toll quality' voice, voice cannot be used as a cushion in this way.

*Fragmentation.* Frame Relay is able to transport voice and data on low bandwidth interfaces as a result of standards-based fragmentation of long data frames so that they do not delay voice traffic.

*SLAs.* Standardized parameters and monitoring methods for QoS and other aspects of SLAs are available, facilitating the transport of voice with the desired level of service.

### 7.3.2 Weaknesses

*Access deployment.* Frame Relay is a successful access technology for business customers, but is not widely deployed in the network backbone and there are no LAN or desktop implementations.

## 7.4  Bandwidth Efficiency

The above strengths and weaknesses of IP, ATM and Frame Relay cover many different business and technological aspects. One aspect on which they can all be compared is the efficiency with which they utilize bandwidth, which depends on the overhead associated with encapsulating the voice packets in the transport data units. We measure efficiency as the number of voice bytes transported as a percentage of the total number of bytes transported.

Figure 7.1 compares the encapsulation discussed for each technology separately in Sections 4.3, 5.3, and 6.4. It shows a voice packet of length *n* bytes inside the overall data structure, plus it shows which parts of the data structure need to be repeated in order to encapsulate voice packets from more than one call. The data structure includes layers 2, 3 and 4 as appropriate.

- For IP, we use version 6. The effect of header compression is shown on the RTP/UDP/

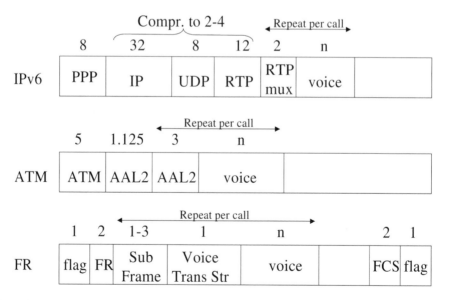

**Figure 7.1**  Encapsulation of voice packets in IP, ATM and Frame Relay, with overhead shown in bytes: not to scale

IP headers and PPP is shown for layer 2. The HDLC flags and FCS are not shown, but the total overhead they contribute is included in the figure of 8 bytes for PPP.

- For ATM, we focus on AAL2 adaptation since it is more comparable to IP and Frame Relay than AAL1.
- For Frame Relay, the HDLC flags and FCS are shown, and do not include fragmentation, Section 6.5.

The efficiency of these technologies evidently depends on the number of voice packets from different calls that are multiplexed within a single transport data unit. Therefore, we present the results in the form of two examples. Table 7.1 is for a single voice packet and Table 7.2 is for ten simultaneous voice packets from different calls. For instance, 25 calls with 60% silence removal would result, on average, in $25 \times 40\% = 10$ simultaneous voice packets.

Tables 7.1 and 7.2 present a comparison of transport technologies for each codec indicated. In order to make this an 'apples for apples' comparison, we assume that the delay introduced by the encapsulation is the same for each transport technology. According to the AAL2 Trunking Specification from the ATM Forum, 1 ms of voice is the basic encoding unit for G.711, G.726 and G.727, whereas the Frame Relay Forum Implementation Agreement uses 5 ms. Therefore, we assume that five samples are used in AAL2 and IP, giving a 5 ms delay for each transport technology for these codecs. The other codecs produce voice packets of different sizes: G.723, 30 ms; G.728, 2.5 ms; and G.729, 10 ms. As we read across Tables 7.1 and 7.2, the delay is the same, but as we read down the tables, we must bear in mind that different voice codecs inherently produce different delays. These tables can therefore be used to answer the question: 'What is the efficiency of each transport technology for codec X?' The tables should be used with care if the reader is addressing a different question such as 'What is the efficiency of each transport technology averaged over all codecs?'

In Tables 7.1 and 7.2, the fixed overhead is the part of the overhead that is not repeated for each voice connection. The variable overhead is repeated for each voice connection (see Figure 7.1). Where Figure 7.1 indicates more than one possibility for the amount of overhead in a field, e.g. for RTP/UDP/IP header compression, an average value is used in Tables 7.1 and 7.2. The figure of 1.125 bytes for the AAL2 fixed overhead is composed of 1 byte of AAL2 overhead plus the 1 byte AAL2 pointer inserted once every eight ATM cells.

The final result of Tables 7.1 and 7.2 is the bandwidth efficiency, calculated as:

efficiency = (number of voice bytes)/(number of voice bytes

+ number of overhead bytes) × 100%

In the case of ATM, the amount of padding required to fill up the ATM cell is added to the overhead.

The efficiency is in general very low when only a single voice packet is being transported and improves significantly when ten voice packets are multiplexed.

The number of bytes of fixed and variable overhead for ATM and Frame Relay is very nearly identical, but the efficiency of ATM is lower due to the need to pad out the 53 byte cells.

Although Table 7.1 shows that Frame Relay is marginally more efficient than IP for a

**Table 7.1** Transmission efficiency of IP, ATM and Frame Relay for major voice codecs with a single call

| CODEC | Voice data unit length | | IP with header compression | | | ATM | | | Frame Relay | | |
|---|---|---|---|---|---|---|---|---|---|---|---|
| | Kbps | Milli-seconds | Bytes | Fixed overhead | Variable overhead | Efficiency (%) | Fixed overhead | Variable overhead | Efficiency (%) | Fixed overhead | Variable overhead | Efficiency (%) |
| G.711 | 64 | 5 | 40 | 11 | 0 | 78.4 | 6.125 | 3 | 75.5 | 6 | 3 | 81.6 |
| G.723 | 6.4 | 30 | 24 | 11 | 0 | 68.6 | 6.125 | 3 | 45.3 | 6 | 3 | 72.7 |
| G.726 | 40 | 5 | 25 | 11 | 0 | 69.4 | 6.125 | 3 | 47.2 | 6 | 3 | 73.5 |
| G.727 | 40 | 5 | 25 | 11 | 0 | 69.4 | 6.125 | 3 | 47.2 | 6 | 3 | 73.5 |
| G.728 | 16 | 2.5 | 5 | 11 | 0 | 31.3 | 6.125 | 3 | 9.4 | 6 | 3 | 35.7 |
| G.729 | 8 | 10 | 10 | 11 | 0 | 47.6 | 6.125 | 3 | 18.9 | 6 | 3 | 52.6 |

**Table 7.2** Transmission efficiency of IP, ATM and Frame Relay for major voice codecs with ten calls

| CODEC | Voice data unit length | | IP with header compression | | | ATM | | | Frame Relay | | |
| | Kbps | Milli-seconds | Bytes | Fixed overheads | Variable overheads | Efficiency (%) | Fixed overheads | Variable overheads | Efficiency (%) | Fixed overheads | Variable overheads | Efficiency (%) |
|---|---|---|---|---|---|---|---|---|---|---|---|---|
| G.711 | 64 | 5 | 40 | 11 | 2 | 92.8 | 6.125 | 3 | 75.5 | 6 | 3 | 91.7 |
| G.723 | 6.4 | 30 | 24 | 11 | 2 | 88.6 | 6.125 | 3 | 75.5 | 6 | 3 | 87.0 |
| G.726 | 40 | 5 | 25 | 11 | 2 | 89.0 | 6.125 | 3 | 78.6 | 6 | 3 | 87.4 |
| G.727 | 40 | 5 | 25 | 11 | 2 | 89.0 | 6.125 | 3 | 78.6 | 6 | 3 | 87.4 |
| G.728 | 16 | 2.5 | 5 | 11 | 2 | 61.7 | 6.125 | 3 | 47.2 | 6 | 3 | 58.1 |
| G.729 | 8 | 10 | 10 | 11 | 2 | 76.3 | 6.125 | 3 | 62.9 | 6 | 3 | 73.5 |

single voice packet, the opposite can be seen in Table 7.2 with ten packets. Frame Relay also has the advantage that although the above calculation is performed for 3 bytes of variable overhead, it is often possible to reduce that to 2 bytes.

If header compression had not been included in the IP calculation, it would have been significantly less efficient. Also the calculation assumes the implementation of RTP multiplexing. If that is not implemented, the efficiency of IP for ten voice packets is exactly the same as for a single voice packet, since each voice packet is put into a separate RTP/UDP/IP data unit. Without RTP multiplexing, the efficiency of IP is given by Table 7.1, independent of the number of voice packets. Thus, RTP multiplexing is necessary in a trunking environment to make IP better than Frame Relay in terms of transport efficiency.

Although IP and Frame Relay compete as transport technologies in an access environment, in the public network backbone the competition is between IP and ATM. It can be seen from Table 7.2 that with RTP multiplexing, IP is more efficient than ATM. Without RTP multiplexing, the efficiency of IP with ten voice packets is given by Table 7.1 and the efficiency of ATM is given by Table 7.2, showing that, for most codecs, ATM is more efficient.

## 7.5 Conclusion

There are not many respects in which apples, oranges and pears can be compared. This chapter has presented one common point of comparison among voice over IP, ATM and Frame Relay, i.e. bandwidth efficiency, plus many individual strengths and weaknesses for each technology as a transport medium for voice. In general, Frame Relay and IP with RTP multiplexing make more efficient use of bandwidth than ATM. In a network backbone implementation, ATM is more efficient than IP without RTP multiplexing. However, the other strengths and weaknesses are often overriding considerations in choosing a transport technology. Most equipment vendors and carriers are hedging their bets by offering all three so as to be sure to suit their customers preferences.

# Part 3

## Broadband Access and Network Evolution

Part 3 of this book shows how some of the technologies described in Parts 1 and 2 can be used in a rapidly growing area of the network: broadband access. We focus on voice over cable and voice over DSL, since they are the two major competing broadband access technologies from the cable companies and the Telcos, respectively. Between them they use the quality of service covered in Chapter 2, the distributed architecture described in Chapter 3, voice over IP from Chapter 4 and voice over ATM from Chapter 5.

Part 3 of this book can therefore be used by

- readers specifically interested in broadband access;
- readers interested in how the technologies described in Chapters 2–5 can be combined in practice.

The book concludes with a chapter describing how the network can be expected to evolve in the future including migration strategies for carriers.

# 8

# Voice Over Cable

## 8.1  Business Issues

The kind of customers who are likely to switch from their ILEC service to a cable company telephone service have been found to be people with moderate to high income currently using several voice features and with moderate to heavy long distance revenues. They also typically have Internet access from a home computer.

The initial market focus for voice over cable is residential multi-tenant buildings. In multi-tenant buildings people move more frequently than people in single family homes. It has been found that when people are moving home, they are also likely to switch their telephone service provider.

An issue with offering telephony over a cable network to apartment buildings is that the wiring inside the building often belongs to the telephone company. In that case if the cable company wants to offer service to such a building, they have to rewire the interior of the building as well as simply providing access to their cable network. Government regulations in many countries require an incumbent carrier to lease their access lines to a competitive carrier, but this applies to the wiring between the customer premises and the first Central Office. It does not usually apply to Telco-owned wiring inside the customer premises.

Another issue with packet voice over a cable network is the lifeline service if the power goes off to the interface in the residence. Residents need to be able to dial emergency services and also to be able to call the electric power company to tell them the power is out. One option is for the Telco to provide lifeline service and for the cable company to provide the second and subsequent lines. However, there is more revenue in the first line than the subsequent lines, so cable companies want to be able to provide all lines including the lifeline service. One way of doing this is to provide battery backup power for the set top box. Another is to deliver power to the residence from the cable company. One cable company supplies emergency power to the residence through power lines which basically follow the cable lines through the residential neighbourhood. They have gas powered generators at the network nodes which have gas piped from the gas company. The generators are not therefore going to fail if the power does go out in the residential neighbourhood.

Figure 8.1 emphasizes the importance of QoS in the packet cable environment. Since

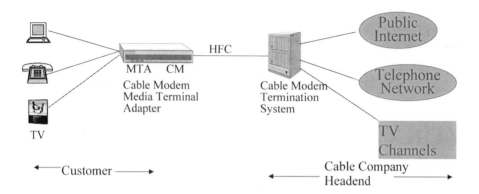

**Figure 8.1**   Providing voice over packet cable

the same network is used for toll quality voice, entertainment quality video and Internet access, adequate QoS must be available for each application.

Another issue with cable modems is that because the coaxial cable network is a shared technology there is the potential for desktop computer users in one household to be receiving packets which were addressed to another household. This way a user can figure out what Web pages their neighbour is accessing. CableLabs Data-Over-Cable Service Interface Specification (DOCSIS) [1] version 1.1 provides security which prevents this intrusion on privacy. It only allows the Internet packets to be read by the person to whom they were originally addressed.

The cable industry has developed sophisticated specifications for packet voice over cable. There are three major specifications:

- DOCSIS [1];
- PacketCable Audio/Video Codecs Specification [2];
- PacketCable Dynamic Quality-of-Service Specification [3].

The major focus of this chapter is on how the packet cable specifications provide telephony. Since they use DOCSIS, we first review those capabilities of DOCSIS that are required for an understanding of packet cable.

## 8.2   DOCSIS

Analogue TV channels on cable networks use 6 MHz channels. DOCSIS is a method of using these 6 MHz channels for digital traffic including data, digital voice and digital video. DOCSIS defines a downstream data rate of 36 Mbps at 6 bits/Hz in a 6 MHz channel. Framing and forward error correction take 9 Mbps out of the 36 Mbps, leaving 27 Mbps for the user. Downstream traffic uses a bandwidth above 54 MHz. Since cable is a shared architecture with multiple users dropping off a single cable, downstream traffic can be picked up by any cable modem. An addressing scheme is used to address packets to specific users. Although the modem is receiving all packets, it only passes on to the user those packets addressed to that user. This raises security concerns, which are addressed by authentication mechanisms built into DOCSIS versions 1.1 and above. These mechanisms use digital certificates based on public key encryption to authenticate each user and are very robust. Telephony packets are typically unicast to a single user, and the addressing scheme also caters for multicast packets, carrying news, sports results, stock prices, etc.

Upstream traffic uses a bandwidth between 5 and 42 MHz, which is more susceptible to noise than the higher frequencies used for downstream traffic. Only 2 bits/Hz are used in order to counteract the effects of noise, and the upstream bandwidth is divided into 1.8 MHz channels with 3.6 Mbps each. Framing and forward error correction consume 50% of this bandwidth leaving 1.8 Mbps for the user.

The total upstream and downstream bandwidths indicated above are shared among several users. For instance, a DOCSIS cable channel capable of 27 Mbps downstream and 1.8 Mbps upstream shared among ten customers will allow each customer to retrieve Web pages at 2.7 Mbps when all customers are simultaneously active. The number of customers to put on a single channel is an engineering and marketing issue, over which cable companies differ considerably.

DOCSIS offers two service types suited to voice:

- *Unsolicited grant.* This service provides $M$ bytes every $T$ seconds, designed for CBR voice, where $M$ and $T$ are chosen according to the voice codec used. It essentially provides dedicated time slots for voice samples, so that there is no contention for bandwidth with data traffic. Early implementations of voice over DOCSIS focussed on 64 Kbps G.711 coding on an unsolicited grant service.
- *Real time polled.* This service is designed for packet voice and will be referred to later in our review of the standard for voice over packet cable.

## 8.3 Reference Architecture

The reference architecture for voice over packet cable is shown in Figure 8.2. The customer has a phone which is connected to the Media Terminal Adapter (MTA), which is connected to a cable modem. In the *business* premises the MTA is usually connected to the

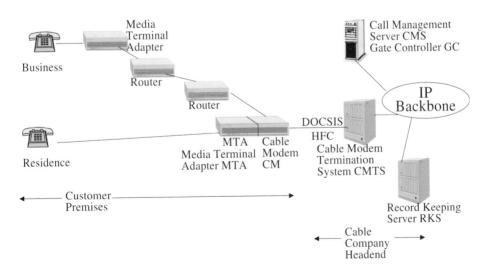

**Figure 8.2** Voice over packet cable reference architecture using HFC, hybrid fibre coax

cable modem via a LAN, whereas in the *residence* the MTA and the cable modem are typically combined into one piece of equipment which is sometimes referred to as the 'set top box'. This cable modem is connected to a coaxial cable network which uses the DOCSIS standard for sending information to the cable company's head end, or remote node, where DOCSIS is terminated by a Cable Modem Termination System (CMTS). The traffic then typically goes over an IP backbone, and hanging off the IP backbone is a Call Management Server (CMS) which usually also incorporates a Gate Controller (GC). The GC is very important to the cable company because it allows them to control what traffic is coming from which residences and it controls access to the IP backbone network. It prevents, for instance, fraudulent use of the IP backbone network. There is also a Record Keeping Server (RKS), which can be used for billing purposes. Let us look at the components of this architecture in turn.

- The MTA is basically a voice gateway. Since an MTA is required for each customer, it is designed with minimal functionality. It does the voice coding and uses an enhanced version of RSVP to set up a QoS-enabled IP connection over the DOCSIS network and it puts the voice packet into the RTP/UDP/IP packet.
- The cable modem implements DOCSIS, that is, it communicates with the CMTS and provides QoS at the physical layer. DOCSIS is essentially an OSI layer 1/2 technology and we're looking at doing voice over IP as layer 3 on top of DOCSIS.
- The CMTS is the interface between the DOCSIS network on the customer side and the IP backbone on the network side (Figure 8.3). The DOCSIS network carries fewer voice calls than the IP backbone, which provides backbone transport for calls from many DOCSIS access networks. Since DiffServ scales better than RSVP (Section 4.4), we use DiffServ on the IP backbone and RSVP on the DOCSIS network. The CMTS interworks between RSVP+ and DiffServ. It also controls access to the IP backbone through the use of gates. Once a call has been admitted to the IP backbone the CMTS polices the amount of traffic the customer is sending.
- The CMS negotiates the session description with the MTA and determines the QoS policy to be used for calls.
- The gateway controller is often implemented in the same equipment as the CMS, although it is a separate logical function. It controls admission of customer traffic to the IP backbone using gates in the CMTS in a master/slave relationship.

The architecture selected by the cable industry for telephony is therefore based on voice over IP. RSVP is used to reserve bandwidth and QoS on the DOCSIS network and DiffServ is used on the IP backbone. When we reviewed RSVP in Section 4.4, we saw that it was originally designed for multicasts over the public Internet, which is an open environment in which customers can request and use bandwidth at will. The packet cable

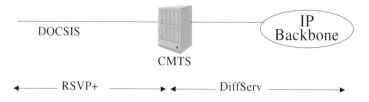

**Figure 8.3**   CMTS

architecture is designed so that the network operator has far tighter control over the use of the network: they can use usage-dependent billing, charging per phone call, and monitor the amount of traffic the customer is sending. Also in IP multicast, the communication is unidirectional by default, whereas for telephony it needs to be bidirectional by default. In order to achieve these objectives an enhanced version of RSVP is used, which we will refer to as RSVP+, and gates are used in the CMTS, controlled by the GC.

## 8.4   Reserved and Committed Resources

In the regular circuit switched PSTN, when we set up a phone call the signalling messages go from one switch to another via the SS7 network, and as the signalling messages arrive at each switch, they reserve time slots on the trunks between the switches for the call. The important thing to note here is that we get those reserved time slots on the trunks before we even start to speak on the call. Before the phone even rings at the other end, we've got a reserved time slot just sitting waiting for us to speak, as and when the phone at the far end finally gets answered. If nobody answers the phone at the far end, all of those time slots were allocated but were never used. The reason for allocating the bandwidth to the customer before the phone rings at the far end is because when the person picks up the phone at the far end, there's only a fraction of a second between them picking up the phone and saying 'Hello'. So there's not time for SS7 messages to work their way back through the network allocating bandwidth to carry that 'Hello'.

In voice over packet cable, reservation of network capacity is done differently. We don't allocate the bandwidth to the customer until we know that the customer is going to start to speak. There's not time for that allocation process in SS7, but over a modern IP network the signalling messages travel and are processed faster and therefore we can wait until the called party actually answers the phone before we allocate the bandwidth in the network. In voice over packet cable we end up with a three stage process involving resources which are 'authorized', 'reserved' or 'committed'.

- *Authorized* means the customer is allowed to use this network. They have signed on as a phone customer of this particular cable company.
- *Reserved* means the customer has dialled a number but we don't know whether the called party is going to answer yet.
- *Committed* means the called party has answered and the phone conversation is taking place.

We go through a two phase procedure when we set up a phone call (Figure 8.4).

- First the customer dials, which is a request to move resources (i.e. bandwidth and processing capacity) from being authorized to being reserved. During this phase, the network determines whether there is sufficient bandwidth for the call. RSVP+ is used to reserve bandwidth and QoS on the DOCSIS network. DiffServ is used to reserve bandwidth and QoS on the IP backbone network. The destination phone rings.
- Phase 2 starts when the called party answers the phone. The GC opens the gates and allows the resources to be committed to the parties on that call. If the customer is to be charged per call, then the RKS starts a billing record for the call. Once backbone resources are committed to the call, the customer traffic is monitored by the CMTS to ensure that it is within the parameters specified when they originally dialled the call.

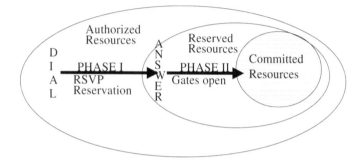

**Figure 8.4** Two phase call establishment

Phase 2 is very important: it prevents fraudulent use of the IP backbone. It must be done fast: in the time it takes the called party to pick up the phone and say 'Hello'. Examples of fraudulent use are:

– customers using statically provisioned voice QoS for public Internet access;
– customers establishing local calls and then changing the destination address in the IP packets to call long distance;
– a called party requesting very high QoS which is billed to the calling party (could be used by customers who do not wish to receive telemarketing calls).

We can renegotiate the bandwidth during the course of a call. Suppose we set up a call which is voice only (Figure 8.5). The calling party is *authorized* to use bandwidth for video and also for voice, but for this call she has *reserved* bandwidth for voice only. The person at the other end has answered the call and therefore the *committed* resources are also for voice only. Half-way through the call the calling party wants to add video (Figure 8.6). The network figures out whether resources are available to be *reserved* for video and finds that they are. Then the called party decides that they can accept video and finally video is included in the *committed* resources.

This idea of committed resources is very useful in efficiently using the bandwidth on the DOCSIS network for Call Waiting. Figure 8.7 shows a call in progress between Bob and

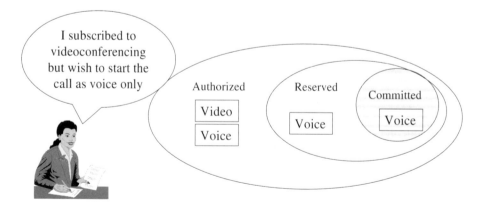

**Figure 8.5** Re-negotiation during call. Step 1: voice only

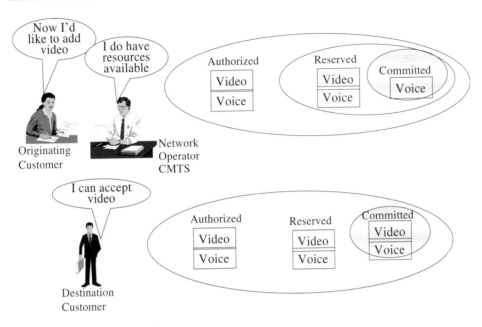

**Figure 8.6**  Step 2: adding video

Sara. Then Jo calls Bob, but Bob's phone is busy because he's already talking to Sara. Resources are *reserved* for two calls: from Bob to Sara and from Jo to Bob. We have *committed* resources on the DOCSIS network for only a single phone call, but then we share those resources between the call to Sara and the call from Jo. That way we only need to *commit* one set of resources and Bob can flip flop between using them for talking to Sara and talking to Jo by manipulating the gates inside the CMTS.

## 8.5  Call Establishment

In this section we run through an example of setting up a call over the packet cable network. Figure 8.8 illustrates Alice calling Bob. The first step is Alice sends out an RSVP+ message from her MTA, which contains the specifications of the call from the

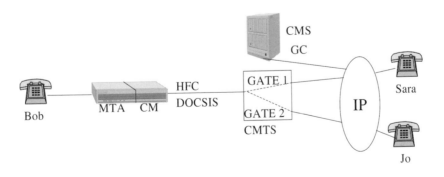

**Figure 8.7**  Call waiting

## - Interception of RSVP messages

**I need to call Bob**

**I'll pretend to be Bob for DOCSIS and backbone reservation**

Alice — MTA CM — HFC DOCSIS — CMTS — IP Backbone — CMTS — CM MTA Bob

**Figure 8.8**   Extensions to RSVP

point of view of the caller. Since we do not use RSVP all the way to the called party, the CMTS in Alice's cable network pretends to be Bob. Bob is actually on a different cable network the other side of an IP backbone. The CMTS does the backbone reservation using DiffServ and also it communicates with Alice and acts as a proxy for Bob from the point of view of setting up the RSVP part of the call. One aspect of this is that it uses RSVP+ to reserve bandwidth in the reverse direction: from Bob to Alice.

The different stages are now described individually at the overview level. For complete detail the reader is referred to Ref. [3].

### 8.5.1   Authorization

Figure 8.9 shows Step 1: authorization. Alice wants to call Bob. The message first goes to Alice's CMS to ask whether she has permission to use the network. Alice is authorized to set up a call. The message also goes to the CMS in the destination network, which verifies that Bob is authorized to receive a call. Also, Bob's equipment confirms that he is prepared to receive a call at this particular time. RSVP+ path messages are sent over the DOCSIS networks from the MTAs at each end. Each message specifies an RSVP TSpec/RSpec for the traffic *in each direction* on the DOCSIS network attached to that MTA. The TSpec/RSpec must be within the parameters authorized for each user. This reservation of *bidirectional* bandwidth is an extension of regular RSVP.

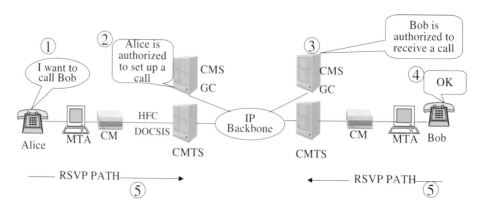

**Figure 8.9**   Call set-up. Step 1: authorization

### 8.5.2   Allocating DOCSIS QoS

The next step (Figure 8.10) is to reserve bandwidth and QoS at the DOCSIS level, which is done by a message from the CMTS to the cable modem. In DOCSIS we can reserve bandwidth in two ways for telephone calls.

- The first is called an unsolicited grant service which is basically a CBR service. It gives us a certain number of bytes at regular time intervals, for instance $M$ bytes every $T$ seconds.
- The second type of service is called real time polled, which is suitable for VBR voice, allowing varying bandwidth to be allocated to the call. The polling interval is determined by the CMTS.

An example of how RSVP parameters are mapped into DOCSIS parameters is given in Figure 8.11. The RSVP parameters are described in Section 4.4, and a summary is as follows:

- TSpec

    - bucket depth: $b$ bytes;
    - bucket rate: $r$ bytes/s;
    - peak rate: $p$ bytes/s;
    - minimum policed unit: $m$ bytes;
    - maximum policed unit: $M$ bytes.

- RSpec

    - reserved rate: $R$ bytes/s;
    - slack term: $S$ s.

The example in Figure 8.11 is for CBR voice, which is identified by the fact that $p = r$ and $M = b$. Therefore, it uses the DOCSIS unsolicited grant service. VBR voice would have been identified if $p > r$ and $b > M$.

Figure 8.11 shows G.711 64 Kbps voice packetized with 80 bytes of voice in an RTP/UDP/IP packet with 40 bytes of header. The header increases the bandwidth required to $r = p = R = 64 \times (80 + 40)/80 = 96\text{Kbps} = 12000$ bytes/s. Since everything is in 120 byte packets, $b = m = M = 120$ bytes. These parameters are mapped onto DOCSIS unsolicited grant service as follows. The grant size is $M$ adjusted for

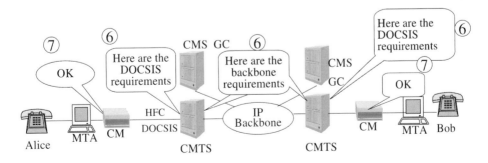

**Figure 8.10**   Call set-up. Step 2: allocating DOCSIS and DiffServ QoS

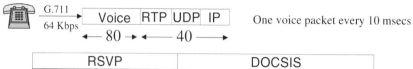

| RSVP | DOCSIS |
|---|---|
| TSpec<br>  b = 120 bytes<br>  r = 12,000 bytes/sec<br>  p = 12,000 bytes/sec<br>  m = 120 bytes<br>  M = 120 bytes<br><br>RSpec<br>  R = 12,000 bytes/sec<br><br>Header suppression 40 bytes | Since r=p and b=M, we use unsolicited grant service, UGS<br><br>Grant size = M + DOCSIS overhead<br>                + ethernet overhead<br>                − header suppression<br>     = 120 + 13 + 18 − 40<br>     = 111<br><br>Grant interval = M/r =10msecs<br><br>Delay = b/r = 10msecs<br><br>If client can accept 15 msecs delay, it sets S= 5 msecs resulting in:<br>Tolerated grant jitter = 5msecs |

**Figure 8.11**   RSVP to DOCSIS example

overhead, which comes to 111 bytes. The grant interval reflects how frequently the users are going to require to send packets of length $M$. They need to send $M$ bytes every $M/r$ s, i.e. the grant interval is $120/12000 = 10$ ms. The delay that will be introduced is $b/r = 10$ ms. If the customer can accept a delay longer than 10 ms on the RSVP connection on one of the DOCSIS networks, they can set a non-zero slack term $S$ in their RESV message. For instance, if they can accept 15 ms delay, they set $S = 15 - 10 = 5$ ms. If this is done it means that DOCSIS has the flexibility to introduce a 'tolerated grant jitter' of up to 5 ms.

### 8.5.3   Allocating DiffServ QoS

At the same time as reserving bandwidth on the DOCSIS network, Figure 8.10 also shows the CMTS sending out a request for DiffServ service over the IP backbone. The policy regarding what DSCP to use is determined by the CMS. Once the bandwidth is reserved on the IP backbone, the CMTS tells the MTA at the customer premises what DiffServ Control Point to use. Figure 8.12 illustrates a decision to use a DSCP of 36. After the call is established (at the end of Section 8.5.5 below) and the caller starts talking over the phone, each voice packet will have a DSCP of 36 put into the IP header by the MTA. When the voice packets arrive at the CMTS the DSCP is checked to make sure that the customer is putting in the right code. This enables the cable network operator to check that the customer is not using better QoS than they first requested. Once this check is performed the voice packet is passed on to the IP backbone.

### 8.5.4   Ringing the Phone

The original RSVP PATH messages sent by the customers in Figure 8.9 have resulted in

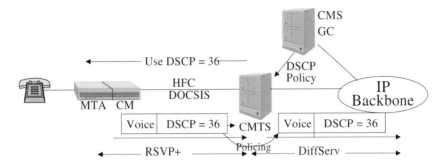

**Figure 8.12**    Interworking with a DiffServ backbone

reservations of bandwidth and QoS in the DOCSIS and IP backbone networks (Figure 8.10). Once these reservations are complete, the CMTSs can respond to the original PATH messages with RESV messages, as shown in Figure 8.13. At this time the CMTSs are again pretending to be the customer at the other end. It therefore appears to Bob's MTA that Alice's MTA has confirmed the connection and vice versa. Bob's MTA therefore causes Bob's phone to ring and Alice's MTA plays a ringing tone to Alice's phone.

### 8.5.5   Committing the Resources

When Bob answers the phone he and Alice expect to be able to start talking. Up to this point the network resources have been *reserved*; now it is time for them to be *committed*. This is done by a 'commit' message which is sent from Bob's MTA to the CMTS, directly Bob's phone goes off-hook (Figure 8.14). Resources are moved from reserved to committed at the DOCSIS level. Bob's MTA also signals Alice's MTA that the call has been answered, so that her MTA can also issue a 'commit' message for her DOCSIS network. The CMTSs signal the RKS to start billing, and the GCs co-ordinate with each other which gates need to be opened. Then they open them, so that the call can access the IP backbone. This entire process is completed within a specified 'post-pick-up delay', i.e. between Bob picking up the phone and saying 'Hello'.

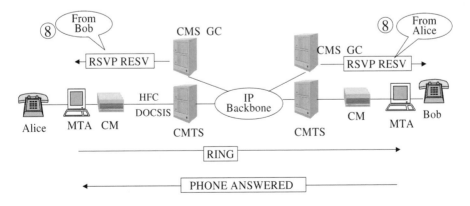

**Figure 8.13**    Call set-up. Step 3: ringing the phone

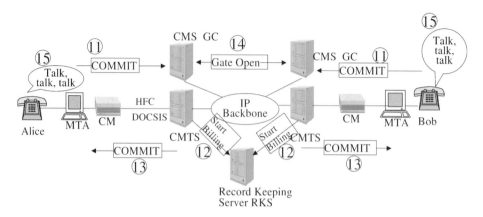

**Figure 8.14**  Call set-up. Step 4: committing resources

## 8.6  Conclusion

Remember the brilliant summer days when the IPers used to have their parades down Main Street? How colourful were their clothes, their flags and banners and how majestically their brass bands proclaimed the virtues of the greatest technology on earth. The simplicity of IP's connectionless paradigm was echoed by the primary colours of those decorations and by the clarity of that music. No connection-oriented pipes had to traverse the IP networks, no call state had to be maintained in the backbone nodes, and no signalling messages had to add to the traffic and the complexity of the software. A click on a Web site in Finland took them effortlessly to a Web site in the Tavoli Islands. And everyone, old and young, men and women, came to watch the show for business and for fun. No matter how you measured the growth of the public Internet, it grew faster than any other technology. Steam engines, electric light bulbs, fax machines, VCRs, and even (so I'm told) the wheel were as nothing by comparison.

And then reality struck: the harshness of commercial profitability had been forgotten in those ebullient days. It was too easy to become an ISP. If one ISP could offer service for $20 a month, another ISP could beat it at $18 and throw in a free pizza to each customer that signed up. It was tough to make a buck from price wars based on flat rate pricing. ISPs needed something else to sell, something that the customer would pay for each time they used it, a few cents a shot, and not notice those cents accumulating into dollars. How about telephony? It can't be difficult, the IPers are the technology whizzes, they must be able to do something the Telcos have been doing for decades. True it may require pipes, more software in some network nodes, signalling messages, billing systems and QoS, but if it will provide a buck on the side, why not go for it.

This chapter has shown how complex IP telephony can turn out to be.

## References

1. CableLabs: Data-Over-Cable Service Interface Specification (DOCSIS).
2. CableLabs: PacketCable Audio/Video Codecs Specification.
3. CableLabs: PacketCable Dynamic Quality-of-Service Specification.

# 9

# Voice Over DSL

In this chapter we first review DSL itself, from the market and technology standpoints. Then we move on to voice over DSL, again from a market and technology perspective.

## 9.1 DSL Market

Analogue telephony has been provided to residential and business customers over a twisted pair of copper wires for decades. T1 is a way of using two such pairs of wires in order to provide full duplex digital communications at 1.536 Mbps. Basic Rate Interface (BRI) is another technology for using the same wires to provide full duplex 144 Kbps. DSL is a way of using the same wires to retain the existing analogue voice channel and supplement it with digital channels. The several flavours of DSL provide a range of digital bandwidth as described in Section 9.2. In this section we analyze the market for DSL.

The business market for DSL includes Small Office/Home Office (SOHO) and Small/Medium Enterprise (SME) access to the public network. Business applications include:

- Internet access;
- corporate network access;
- public ATM network access;
- telephony access.

The residential market for DSL includes:

- Internet access;
- local content access;
- local multimedia conferencing;
- telephony access;
- corporate network access for telecommuters.

Figure 9.1 illustrates a typical situation of a voice call over DSL using the analogue channel in the DSL line, plus the DSL line has got a lot of digital bandwidth which is typically used for data. That way we can have an analogue phone call and Internet access over the DSL line simultaneously. 'Voice over DSL' is one step further than that, where we take other phone calls and instead of sending them over the analogue channel in the DSL line, we send them over the digital channel in the DSL line. We can get several phone calls simultaneously (one analogue and the others digital) plus the data traffic all over a single DSL line.

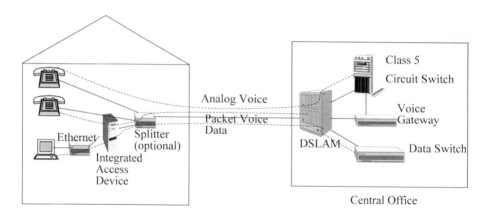

**Figure 9.1**   Analogue and packet voice over DSL

A single DSL link can be used for multiple applications at the same time. For instance, in a residential situation, four family members can simultaneously use analogue and digital telephony, Internet access and watching a local sports team using video on demand.

Figure 9.1 illustrates a splitter at the customer premises which separates the analogue voice channel from the digital channels. Splitterless DSL is also available, with the same analogue voice channel but a reduced digital bandwidth. Splitters must be installed by the Telco at the demarcation point between the customer and the carrier. A truck roll can therefore be saved by using splitterless DSL. Alternatively, microfilters can be used for each telephone which can be installed by the customer.

## 9.2   DSL Technology

Behind the (optional) splitter in the customer premises is an IAD, whose functionality is to code and packetize the voice and set up voice calls. On the network side the IAD incorporates an ADSL Terminal Unit-Remote (ATU-R), with an RJ45 port. On the customer side, it can have ports for analogue, RJ11, digital phone lines, and wireless phones, plus ethernet, usually RJ45 or wireless, IEEE 1394 and/or USB ports for data/audio/video equipment. The IAD acts as a media gateway and may implement MEGACO (Section 3.3). So starting in the customer premises, we take some voice, we code it and send it over the DSL digital bandwidth to the DSLAM which passes it on to a voice gateway which decodes it and sends it as G.711 to the circuit switch. Alternatively, the DSLAM sends it over a packet network to a remote voice gateway. The data traffic from the customer premises follows the same path as the voice traffic to the DSLAM and is then sent over the long haul packet network.

### 9.2.1   Physical Layer

DSL provides raw bandwidth at the physical layer. There are many flavours of DSL.

- Full Rate ADSL is the most well defined at the time of writing. It uses the spectrum above 26 KHz to provide multiple 32 Kbps channels [1] over 18 000 feet (5.5 km).

Each channel uses a 'tone', and tones are spaced every 4.3125 KHz. The use of multiple tones of this type is referred to as Discrete MultiTone (DMT). The highest frequency used for a tone is 1.1 MHz, which gives a total bandwidth of 8.1 Mbps downstream. The upstream bandwidth is 0.8 Mbps. Customers not requiring the full bandwidth can be configured to a lower bandwidth by the service provider not activating some of the higher frequency multitones, even though they are physically available in the equipment.

- ADSL Lite provides 1.5 Mbps downstream and 0.5 Mbps upstream [2].
- Symmetrical Digital Subscriber Line (SDSL) provides 0.768 Mbps in each direction over 15 000 feet (4.5 km).
- SDSL2 [3] provides 2.36 Mbps in each direction.
- High Speed Digital Subscriber Line (HDSL) [6] provides 1.5 Mbps in each direction over two twisted pairs.
- HDSL2 provides 1.5 Mbps in each direction over one twisted pair.
- Very High Speed Digital Subscriber Line (VDSL) provides downstream rates of 12.96 Mbps over 4500 feet (1.4 km), 25.82 Mbps over 3000 feet (0.9 km) and 51.84 Mbps over 1000 feet (0.3 km), and upstream rates from 1.6 to 2.3 Mbps.

In North America 70% of local loops qualify for ADSL implementation, i.e. they are less than 18 000 feet, and satisfy a number of wiring specifications relating to the gauge of wire and the number of bridged taps. In Europe, loop lengths tend to be shorter than in North America resulting in a greater potential availability of DSL.

With the emergence of the multiple flavours of DSL listed above, the question arises as to which flavour can be implemented on a given local loop. This issue is addressed by Ref. [4], which is under development at the time of writing.

In order to provide DSL service to a customer located more than 18 000 feet from the Central Office (CO), the service provider can install a remote DSLAM part way between the customer and the CO. The remote DSLAM aggregates traffic from a number of DSL lines (each less than 18 000 feet long) into a SONET transmission facility to the CO.

### 9.2.2   QoS

In ADSL there are two types of channels. One of them (the interleaved path) uses interleaving, which is described in Section 2.2.2 as being a method of error control which has the disadvantage that it introduces some delay. Interleaving increases the effectiveness of Forward Error Correction. The other type of channel (the fast path) in ADSL doesn't use interleaving; therefore, its not so good at error control, but it introduces less delay. Therefore, in doing voice over ADSL we need to make a choice between having a tight end-to-end delay and having good error control. You can't have low delay and also low loss ratio over DSL. Its a trade-off between one or the other.

So the interleaved path which gives us the low error ratio and the non-interleaved path which gives us the low delay can be regarded as different subsets of DSL with a partition between them. The interleaved path is suited to data traffic which is sensitive to errors but not to delay. For voice, some carriers may also choose to use the interleaved path so long as total end-to-end delay is not too great. Others may choose the fast path so as to reduce delay. ADSL equipment can therefore be configured with a specified number of 32 Kbps channels on the fast path and other channels on the interleaved path. Also, as voice calls are set up and torn down, we need to be able to dynamically change this partition between the fast path and the interleaved path.

ADSL Lite only offers the interleaved path.

## 9.3   The Role of ATM in DSL

DSL is a physical layer technology providing 32 Kbps channels that can be aggregated into multiples of 32 Kbps. However, the control over these channels is limited. The DSL Forum has therefore recommended the use of ATM on top of DSL. ATM provides the layer 2 capabilities that harness the raw bandwidth and makes it into useful pipes. The advantages of combining ATM with DSL are:

- ATM is a switched technology so that ATM virtual circuits do not have to terminate at the ATU-R or ADSL Terminal Unit-Central Office (ATU-C), but can be switched through ATM networks at either end. At the business customer premises this provides communications over an ATM LAN. At the public network ATU-C, this provides communications over an ATM/DSL access onto a long haul ATM network.
- ATM provides SVCs so that the customer can set up connections themselves, and can, for instance, access the corporate network during the day for telecommuting, and the Internet in the evening for entertainment. This saves the network operator having to configure connections for the customer.
- ATM allows different customers to communicate with the same destination via different routes. This is important for public Internet access. If customer #1 is subscribed to ISP #1 and customer #2 is subscribed to ISP #2, and each customer wants to visit www.altavista.com, it is important that each customer's traffic gets to that common destination via different ISPs. This is tough to do if the routing decisions are all made at the IP layer. ATM can be used to provide a virtual circuit from customer #1's ATU-R to ISP #1, and another virtual circuit from customer #2's ATU-R to ISP #2.
- ATM provides different QoS on each connection, so that it can be used to reduce delay on voice traffic when data is also traversing the same DSL link.

## 9.4   Voice Over DSL Applications

### 9.4.1   CLEC and ILEC Applications

A service provider can make some money by using DSL to provide Internet access and also providing one analogue phone call at the same time. But they can't make a lot of money if they limit it to just one phone call. The amount of revenue which a service provider gets from voice traffic is typically higher than the amount of revenue from the same bandwidth of data traffic. So a service provider can make money out of DSL:

- by offering Internet access to residential subscribers, from which they get a small amount of revenue;
- but they can make more money out of selling DSL to small businesses who want more than one simultaneous phone call. Then they can get the revenue from all of those phone calls in addition to the revenue from the data traffic. From the point of view of

selling DSL digital bandwidth the service provider can expect more revenue from phone calls than from Internet access.

The voice over DSL market is therefore the Small/Medium Enterprise (SME) or the branch office of a large corporation as opposed to the residential user. The market is a situation where, pre-DSL, the local voice expenditure is higher than the local data expenditure. In other words, its a small business with say half a dozen phone lines and not much data traffic. That kind of customer is the customer that service providers are focusing on for the voice over DSL service. It presents an opportunity for the customer to reduce their local voice costs which they want to do because their voice traffic costs more than their data traffic. It also gives an opportunity for the carrier to obtain significantly more revenue than by offering data over DSL alone, because the revenue from the voice calls is more than the revenue from the data traffic.

The ILECs are not the first to offer voice over DSL, since they stand to lose some of their existing voice revenue when the customer switches from multiple analogue lines to voice over DSL. The CLECs, however, are the prime movers. Even when they offer voice over DSL at a lower price than the ILEC analogue voice service, they still make more revenue than by offering data only. It is thus competition from the CLECs that forces the ILECs to start offering voice over DSL.

Another reason for focusing on the SME market is that there is less competition from the cable companies, which have deployed their networks primarily in residential areas because their initial market focus was entertainment video. Cable networks do not typically pass commercial office parks where many small businesses are located.

An area of voice over broadband access in which cable companies and Telcos compete is residential apartment buildings. In this case, both carriers have access networks to the building. In some cases the Telco also owns the telephone wiring inside the building, giving it an advantage over the cable company from the viewpoint of providing a telephone service to tenants.

Other advantages of voice over DSL from the viewpoint of the incumbent carrier are that it can be used to provide additional phone lines to customers in areas where there are no more spare twisted copper pairs and it can also be used to offer additional lines at discounted prices so as to compete with cable companies.

Voice over DSL is particularly attractive to CLECs. In the USA the FCC has stipulated that incumbent carriers must lease parts of their access networks to CLECs 'at cost'. In the case of a twisted pair access line owned by an ILEC, the ILEC incurred a cost to install that line in order to offer analogue telephony. If the CLEC does not want to offer analogue telephony, but only digital telephony, then arguably there is no 'cost' to the ILEC in leasing that capability to the CLEC. The ILEC retains its revenue from analogue telephony to offset its cost of installing the twisted pair line. The higher frequency capability of the twisted pair was always there, but the ILEC chose not to use it. CLECs are therefore able to offer voice over DSL at zero cost to the ILEC, and therefore do not need to pay the ILEC a lease. The only cost to the CLEC is the cost of the DSL terminating equipment in the customer premises and collocated in the ILEC CO.

As described in Section 9.2, there are many different flavours of DSL. In order to decide which is appropriate for voice, a primary consideration is whether the bandwidth requirement is symmetrical. Interactive telephony applications require the same bandwidth in each direction, i.e. the bandwidth requirement is symmetric, whereas many data applications, such as Web browsing, require more downstream bandwidth than upstream. Figure 9.2 shows three options. If we've only got voice traffic, then we need SDSL because voice traffic is symmetric. If we've got a small office or a home office (SOHO), then typically we need ADSL because we've got a small amount of voice, but a lot of data traffic accessing the corporate network or accessing the public Internet where typically the bandwidth requirement is asymmetric. In the SME or branch office market we typically need SDSL because there are several phone calls which have a symmetric bandwidth requirement and our data traffic is also fairly symmetric because we may have some servers located in the customer premises as well as remote servers. HDSL and HDSL2 also provide symmetric bandwidth.

That is basically the business case for putting voice over the digital channels in DSL. We now distinguish three major applications of voice over DSL:

- loop emulation, which makes the voice over DSL channel look like an analogue access line from the viewpoint of the network and terminates it on a class 5 circuit switch;
- PBX extension, which provides telecommuter access to a corporate voice network;
- packet voice network access, which provides access to a long distance carrier which is also transporting voice over their packet network.

### 9.4.2   Loop Emulation

Loop emulation for the incumbent local carrier (Figure 9.3) involves taking the phone traffic, digitizing it in the IAD, bringing it into a CO over DSL and then decoding it and sending it straight to the class 5 circuit switch. This is loop emulation in the sense that from the point of view of the circuit switch, it looks as though there is an analogue link over a twisted pair of wires going all to the way to the customer's phone. Calls are therefore set up using the regular dialling system and routing systems in the circuit switched PSTN.

Loop emulation for the competitive local carrier is slightly different (Figure 9.4). The voice goes through the IAD and is digitized and sent on the digital channel of the DSL to a DSLAM which is located in the ILEC's CO. This DSLAM may belong to the CLEC or it may be leased from the ILEC. The traffic then goes over a packet network to the CLEC's point of presence where it may, for instance, go into a voice gateway and into the CLEC circuit switched network at that point.

The technology for loop emulation is described in Section 9.5.3.

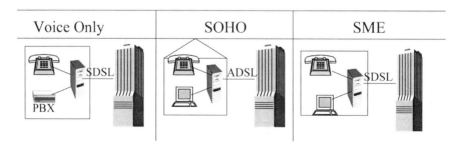

**Figure 9.2**   Selection of DSL to suit the market

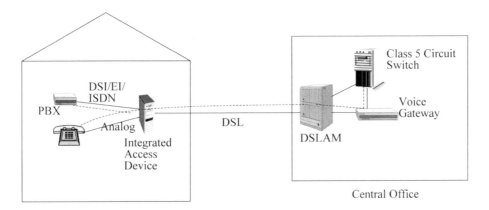

**Figure 9.3**   Loop emulation for the incumbent local carrier

### 9.4.3   PBX Extension

The next example application is PBX extension (Figure 9.5), where a PBX in the corporate network at a corporate site provides services, call control features and call set-up to an analogue phone in a telecommuter's house. The traffic goes over DSL through the DSLAM in the ILEC CO and over a packet access network. The diagram then illustrates two options:

- The traffic could go over a corporate packet network to a MEGACO call controller acting as a PBX. The PBX features are provided by software on the call controller, or on a separate feature server as described in Section 3.2. This option is to be used in the case where the corporate network is a packet network for voice.
- Alternatively, in order to reach a physical PBX the voice would pass through a gateway in order to be converted to G.711. Internal calls over a corporate circuit network could then be made via this PBX. This option is used in the case where the corporate network is a circuit network.

### 9.4.4   Public Packet Network Access

For public packet voice network access the traffic goes over the DSL line through the

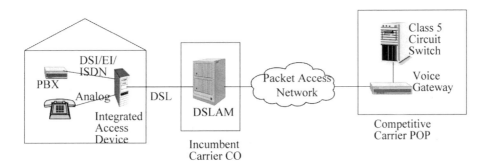

**Figure 9.4**   Loop emulation for the competitive local carrier

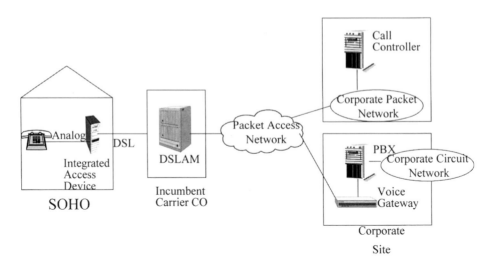

**Figure 9.5** PBX extension

DSLAM and over a long distance packet network before it hits a voice gateway which decodes the voice and sends it to a class 5 switch. Notice that in this case, the voice traffic goes into the class 5 switch through a trunk interface because this is the long distance part of the network, whereas in the case of loop emulation we had the voice traffic going into a line side interface on the class 5 switch.

### 9.4.5  Lifeline Support

The IAD requires power to operate, so the issue arises as to how telephone service can be provided in the event of a power failure at the customer premises. A traditional POTS phone works when the power is out, and it is important to provide this 'lifeline support' in voice over DSL. There are four options.

(a) Use a separate twisted pair for an analogue phone which does not require power from the customer premises.
(b) Use the analogue channel in ADSL which is maintained even when power is out at the IAD.
(c) Use battery backup for the IAD.
(d) Configure the IAD so that if power fails, all voice over DSL phones become extensions to the analogue line.

Customers will pay more for their first phone line, i.e. the line with lifeline support, than for subsequent lines. Many CLECs therefore want to provide all the phone lines, not just the second and subsequent lines. They would tend to be in favour of options (c) and (d) above. ILECs, wanting to retain the primary line, even when a CLEC is offering DSL to the same customer, would prefer options (a) and (b).

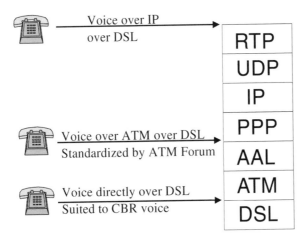

**Figure 9.6**    Voice can enter the protocol stack at three levels

## 9.5    Voice Over DSL Technology

The voice can get over the DSL at different levels in the protocol stack (Figure 9.6). It could be voice over IP over DSL or voice over ATM over DSL or voice directly over DSL. The last of these three is particularly suited to constant bit rate voice and is not often used. The one which is most often used is voice over ATM over DSL which is standardized by the ATM Forum.

### 9.5.1    Voice Over IP Over DSL

Voice over IP over DSL is suited to the situation where the customer premises equipment uses IP telephony (Figure 9.7). The voice packets are mapped into RTP, then UDP, then IP as described in Section 4.3 by existing customer premises equipment. The IAD is not involved in voice coding or packetization since those functions have already been done by the time the IP packets arrive at the IAD. For instance, if we have an IP PBX at the

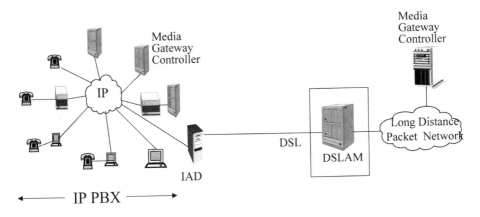

**Figure 9.7**    Voice over IP over DSL

customer premises, then it makes sense to keep the voice in IP format over the DSL access line.

Voice over IP over DSL is particularly suited to service providers offering voice over IP in the long haul network. This enables seamless transport of voice in IP format all the way from the customer premises voice over IP device, through the access line and intercity network.

Figure 9.7 illustrates two MGCs, one in the customer premises and one in the public network. Either or both could be used to set up calls and to provide features either from software on the MGC or on a separate server.

Section 9.3 above has indicated various advantages of having ATM over DSL and therefore the voice over IP traffic is likely to be transported over ATM using AAL5 in order to get over the DSL line, following the protocol stack in Figure 9.6. There are various ways of doing this including 'Classical IP over ATM' RFC 1483/1577, and MPLS. These may be implemented in the IAD in order to carry IP data traffic over ATM over the DSL access line, so that there is no additional cost for transporting voice in the same way.

### 9.5.2  Voice Over ATM Over DSL

Alternatively, if we don't have voice in IP format at the customer premises, then we can send voice over the DSL line by putting it into ATM format (Figure 9.8). In this case, the IAD codes the voice and puts it into AAL2 packets which are fed into ATM virtual circuits and sent over the DSL access line.

The IAD can operate in two different ways.

- It can act as a cross connect so that each voice port always gets mapped into the same AAL2 channel.
- It can act as a concentrator where the channels are allocated dynamically.

When the voice reaches the public network DSLAM it can be handled in three alternative ways.

- It can be sent over another packet network, e.g. an IP network. In this case the voice packets need to be extracted from AAL2 and mapped into RTP and UDP and IP.
- It can be sent over a public or private ATM network, involving seamless operation, maintaining the voice packets in their original AAL2 format.

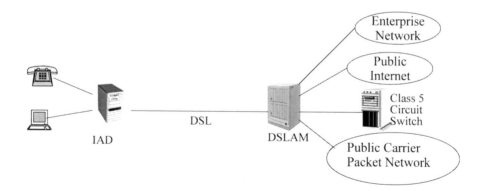

**Figure 9.8**  Voice over IP may also use ATM

- It can be sent via a voice gateway to a class 5 circuit switch. The voice gateway extracts the voice from AAL2 and re-codes it into G.711 for the circuit switch. The use of AAL2 on a DSL access line is very similar to the use of the AAL2 for trunking (see Section 5.4.4). Here we're using AAL2 in the access network, but essentially we're using it in the same kind of trunking situation, in the sense that there is a non-ATM part of the network at either side: in the customer premises and in the class 5 circuit switch. So the standard which we use for doing voice over AAL2 in the access network is very similar to the standard which we use for voice over ATM in the backbone or long haul part of the network. This option includes loop emulation (Section 9.4.2), which we now examine in more detail.

### 9.5.3   Loop Emulation Service

The ATM Forum has specified a voice over ATM over DSL 'Loop Emulation Service (LES)' [5], which can be used by an incumbent or by a competitive local carrier to make voice over DSL channels 'look like' a twisted pair of copper wires from the point of view of the class 5 circuit switch (Figure 9.9). The switch takes in signalling messages and voice as if they had come from a phone on an analogue line, whereas in fact they have been coded digitally and sent over ATM over DSL. This allows existing class 5 switches to provide service transparently to voice over DSL customers.

LES is based on AAL2 and has some commonality with AAL2 trunking (see Section 5.4.4). There are five main differences between AAL2 trunking and LES.

- In AAL2 trunking, signalling messages may be interpreted by the ingress gateway so as to set up connections across the ATM backbone accordingly. In LES, the signalling messages from the customer equipment are passed transparently through to the network-based voice gateway so that they can be passed on to the class 5 switch as though they had come from the customer's phone. No routing decision is made by the customer premises IAD. Its ATM virtual circuits for voice are connected permanently to the class 5 switch. CAS is carried in data packets on the same AAL2 channel as the voice that they are associated with. CCS for all voice channels is carried over AAL2 using channel identifier, CID = 8.
- AAL2 trunking involves peer-to-peer messaging between two gateways at either side of an ATM backbone network. LES is more of a master/slave relationship with the network-based voice gateway controlling the customer premises IAD.

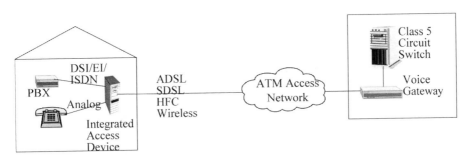

**Figure 9.9**   Voice over ATM over DSL. ATM Forum LES

- In standard AAL2, AAL2 data units can overlap between two ATM cells. LES includes a 'Lite' option in which each ATM cell only contains a single AAL2 data unit. This simplifies implementation and hence reduces the cost of the customer premises IAD at the expense of low DSL bandwidth utilization. More expensive equipment can implement fully fledged AAL2.
- AAL2 channels can be assigned explicitly, in which case the network gateway uses CID = 8 to dynamically allocate customer premises ports to specific CIDs. Alternatively, customer premises ports can be statically assigned to CIDs within the following ranges:

  - analogue phones, CID range 16–127;
  - ISDN D-channels, CID range 128–159;
  - ISDN B-channels, CID range 160–223.

- LES has an embedded operations channel, CID = 9, which is used to send SNMP messages. The network-based gateway implements SNMP software which can query the status of the customer premises IAD.

### 9.5.4   Voice Directly Over DSL

If we do voice directly over DSL, we don't benefit from silence removal or packetization and the voice just stays in constant bit rate format as it is transported over the DSL physical layer in 32 Kbps channels. This option is simple to implement since it does not use any of the higher layers in the protocol stack, but then it does not benefit from the features provided by those layers, for instance the advantages of ATM, described in Section 9.3.

### 9.5.5   Voice QoS

As described in Section 9.2.2, ADSL offers a 'fast path' with low delay and an 'interleaved path' with low error ratio. The QoS provided by IP or ATM is limited by the capabilities of the physical transmission system. Although both DiffServ and ATM allow the user to request specific delay/error characteristics of an end-to-end connection, a very low delay and very low error combination may simply not be available on an ADSL link. The user will receive a signalling response 'QoS unavailable'. IP and ATM signalling need to be able to:

- select either the interleaved or fast paths from ADSL;
- dynamically change the partition of bandwidth between the interleaved and fast paths.

## 9.6   Voice Over DSL Capacity

Section 5.3.3 indicates that G.729 voice at 8 Kbps over AAL2 results on average in 36.1 bytes of voice and 16.8 bytes of overhead per ATM cell. Thus, the original 8 Kbps of voice becomes $8 \times 53/36.1 = 11.75$ Kbps including the ATM and AAL2 overhead.

Since SDSL provides 0.768 Mbps, it has a maximum capacity of $768/(11.75 \times 2) = 32$ simultaneous calls assuming a 50% utilization of the DSL transmission facility.

The ADSL upstream bandwidth of 0.8 Mbps can similarly be used for 34 calls at 50% utilization.

It is important not to plan on using 100% of the DSL bandwidth for packet transport. The burstiness of packet traffic, whether voice or data, requires spare capacity to cope with the statistical multiplexing effects of several bursts occurring nearly simultaneously. Even so it can be seen from the above calculations that the combined effects of voice compression, packetization, statistical multiplexing and transport overhead allow more calls to be carried on a single twisted pair than over the two twisted pairs required for 24 voice calls over T1, and at considerably reduced equipment cost.

## 9.7   Conclusion

In 1989 the ITU created the seeds of ATM, and cast them on a pilotless sailing ship on a becalmed ocean, with no wind to blow her to any shore, and no captain to choose which shore to steer for. Soon a storm arose, hurling the ship to the land of LAN switching. It was a fertile shore on which many plants already grew, ethernet, token ring and FDDI, but there was space for ATM to take root, even though it was ultimately to be displaced by gigabit ethernet. Meantime, the ATM Forum had been preparing another landing place, the end station interface, and caused an eddy to grasp the ship from its mooring and lead it to this new shore. But it proved a rocky shore, and ATM had little to sustain itself in competition with the established undergrowth of ethernet, USB and IEEE1394.

The Telcos and ISPs did not like to see valuable ATM seeds being tossed this way and that, so they sent a captain to grasp the helm of the ship and steer it to their farms in the land of the public network backbone. Here the seeds flourished as never before. They had no natural enemy on this fertile soil and spread across the land. Each year the size of the ATM backbone crops doubled until it spread to the banks of a broad river. On the other side were two dinosaurs: terabit routing and dense wavelength division multiplexing. Whenever an ATM seed blew across the river and took root on the other side, one of the dinosaurs gobbled up the young plant. Then the dinosaurs found a way across the river and started feeding on the ATM crops, pushing them back so that the only place left for ATM to grow was the land of the access network. The Telcos and ISPs fenced the land in, keeping out the dinosaurs, and provided DSL fertilizer to help ATM to thrive. Fertilizer is of little use without seeds, and seeds grow poorly without fertilizer. The duo became successful and ATM covered the rolling hillsides, under the watchful eye of its public network farmers, who monitored the growth of packet cable ranchers on the other side of the valleys with cautious eyes.

Has ATM at last found its holy grail, or will the dinosaurs one day scale the fences and invade the access network farmland with fibre to the home at last? What about the ranchers on the other side of the valleys? Will their cattle one day come stomping over the ATM crops plundering all in their path?

One thing for sure is that there will always be change. But for now, ATM and DSL are a prolific duo for voice, data and video.

## References

1.  ITU-T G.992.1 ADSL Full Rate 8.1 Mbps Downstream, 0.8 Mbps Upstream.
2.  ITU-T G.992.2 Splitterless ADSL 1.5 Mbps Downstream, 0.5 Mbps Upstream.
3.  ITU-T G.991.2 SDSL 2.36 Mbps Bidirectional.

# 10

# Network Evolution

In this final chapter we briefly review some issues associated with directions in which the telecommunications networks of the world are currently evolving, including wireless networking, DWDM, and MPLS, and how these developments impact packet transport of voice. We conclude by examining migration strategies for carriers to transfer voice from circuit to packet networks.

## 10.1  Wireless Networking

In all wireless networks, bandwidth is at a premium, so their proliferation increases the demand for packet voice, which saves bandwidth by silence removal.

### 10.1.1  Satellite and Aircraft-Based Systems

Voice is currently transported over the following wireless interfaces:

- Geostationary satellites have been used within the public network for decades to transport international telephone traffic. Their advantage derives primarily from the simplicity of their operation. Since each satellite has a footprint of over a third of the earth's surface, most international calls can be handled by a single satellite hop with no need for inter-satellite communication. Since the satellites stay fixed relative to the earth, there is no need to hand off calls from one satellite to another. The downside to geostationary satellite communication for voice traffic is the half second round trip delay which limits interactivity and is way above the generally acceptable delay of 100 ms discussed in Section 2.3. The proliferation of fibre, particularly under the oceans, has therefore resulted in a migration of telephony traffic from satellite to terrestrial links.
- Low/Medium Earth Orbit satellites (LEOs and MEOs) a few hundred miles above the surface of the earth solve the problem of delay. They have been deployed at a time when the complexity of handing off calls from one satellite to another is something engineers take in their stride based on experience from cellular networks. The satellites orbit in North/South planes (as opposed to the geostationary equatorial orbit, Figure 10.1) and as an earth station sees its satellite dip down towards the Southern horizon, another satellite is rising in the North onto which the call is automatically handed off. On-board switches, many based on ATM technology, facilitate the switching of packet-

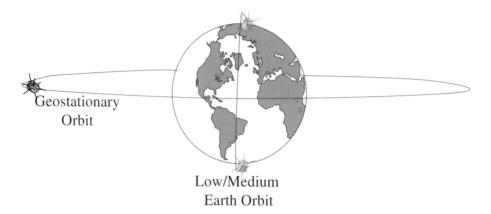

Geostationary
Orbit

Low/Medium
Earth Orbit

**Figure 10.1**   Satellite orbits: not to scale

based calls from earth station to earth station and from satellite to satellite. Inter-satellite switching means that the entire telecommunications infrastructure is in the sky instead of being on the ground. The voice transported over LEO/MEO systems is typically packet switched. The first major limitation of LEO/MEO systems is the bandwidth of the up/down links. Although voice requires only a few Kbps, the business case for the multi-billion dollar capital investment in the many satellites required (around 100 satellites, e.g. ten in ten North/South orbits) is dependent on capturing some of the market for Internet access, for which customers continue to expect higher and higher bandwidth. The second limitation is the power requirements of the earth stations that render them far less portable than cellular phones. The third limitation is that it is impossible to design a system that gives less than global coverage. Since the earth rotates West/East underneath the North/South orbits of the satellites, multiple satellite orbits are required to provide continuous coverage over a single part of the earth's surface as it rotates. Even if a network operator wished to provide service only in, say, Europe, they would need satellites in orbits all around the earth, waiting for Europe to arrive on its daily journey. The network operator needs just as many satellites (and therefore capital investment) to offer service for Europe as to offer global service. Whereas many innovative technologies have originally been deployed by small companies, LEO/MEO systems are the province of only very large companies that can afford to invest in a global infrastructure.

- Aircraft-based systems allow small operators to offer wireless service in a single city. Two aircraft (for redundancy) with ATM switches, circling at about 50 000 feet, can be used to provide metropolitan service, with a third aircraft waiting to take over when one of the others needs re-fuelling. Hand-off of calls from one switch to another is performed when one aircraft replaces another. The main application of such a system is enterprise voice and data networking based on fixed earth stations with dishes less than a foot in diameter mounted on the roofs of office buildings. In order to provide service across a metropolitan area, the bandwidth per customer is at a premium, and packet switching of voice is used to reduce bandwidth requirements.

## 10.1.2   Wireless ATM

Terrestrial cellular networks have seen very rapid growth over the past few years as governments have made additional spectrum available, as packet voice technology has become portable and as the demand for wireless Internet access and compressed Web content has developed. We have seen in Chapter 5 how the packet voice arriving over the wireless interface at a cellular company base station has given rise to a demand for AAL2 networking on the wired network between the base station and the switching centre. In this section we address the question of how ATM connections are managed when a mobile user moves among base stations.

A common misapprehension about the ATM Forum's 'Wireless ATM' (WATM) speci-fication [1] is that it deals entirely with how to send ATM cells over a wireless interface. Although it does cover that point, a major additional contribution is how ATM signalling can be extended in the wired portion of a cellular company network to keep in touch with a mobile terminal as it moves from cell to cell. Examples of a mobile terminal are a mobile phone, personal digital assistant, or laptop computer. Voice is often a major medium of communication, and the WATM specification allows QoS parameters to be specified for voice and/or other media, such as data and video.

In this section, we are using the word 'cell' with two different meanings: (i) an ATM cell, i.e. a 53 byte data unit; (ii) a geographical region of a cellular network served over a wireless interface from a single base station. The context should avoid confusion between these totally different concepts.

Figure 10.2 illustrates the reference architecture, with three cells served by three base stations or points of attachment. The mobile terminal communicates over the wireless interface with the Point of Attachment (PA) of the cell in which it is currently located. That PA communicates over a wired ATM network, either with the destination or with a point at

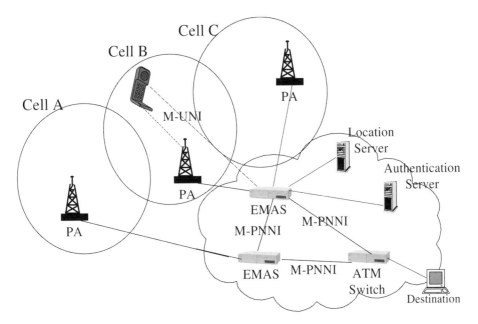

**Figure 10.2**   WATM reference architecture

which the call enters the circuit switched PSTN. The ATM network may contain regular ATM switches, and also needs special functionality in other ATM switches in order to maintain communication with the user as they move from cell to cell. Switches with this additional capability are referred to as End-User Mobility ATM Switches (EMASs). Communication between the mobile terminal and the nearest EMAS is over a Mobile User-to-Network Interface (M-UNI).

Figure 10.2 also shows that an extension to PNNI, known as Mobile Private Network-to-Network Interface (M-PNNI), is used between EMASs in the wired ATM network. M-PNNI contains extensions to regular PNNI messages that pass transparently through regular ATM switches.

The final elements in the WATM architecture are:

- the location server, which maintains a database that can be used to find a specific end-user system. The database includes the permanent ATM end system address of the mobile phone, and the ATM address of the EMAS with which it is currently communicating. The location server is used when a call is being set up *to* a mobile terminal, whereas Figure 10.2 refers to a call *from* a mobile terminal;
- the authentication server, which authenticates mobile users as part of the process of giving them permission to use the network.

As the mobile terminal moves from one cell to another, the increased distance from its original PA results in a degradation of the wireless signal. The mobile terminal needs to start to communicate with a new PA so that a new ATM SVC or SVCs may need to be established in order to maintain communication across the ATM network. The mobile terminal may have multiple connections in place at any one time, all of which require hand-off. Figure 10.2 illustrates two situations.

- *Intra-EMAS hand-off.* When the mobile user moves from cell B to cell C, the new PA is connected to the same EMAS as the original PA, so that maintaining communication with the end user can be accomplished within that single EMAS.
- *Inter-EMAS hand-off.* When the mobile user moves from cell B to cell A, the new PA is connected to a different EMAS than the original PA, so that maintaining communication with the end user requires signalling between these two EMASs.

The hand-off process can be initiated either by the current PA or by the new PA. For the sake of illustration we now describe the former of those two options which consists of the following sequence of steps.

1. When the radio layer in the mobile terminal detects degradation of the wireless signal from the current PA below an acceptable threshold, it searches for other PAs with which it is able to communicate and passes a prioritized list of possible PAs to the ATM layer in the mobile terminal.
2. The ATM layer in the mobile terminal uses the M-UNI to pass the prioritized list of PAs to the current EMAS.
3. The current EMAS looks at the list of PAs to see whether it can use a PA managed by itself which has sufficient capacity to handle the connection(s). It also sends messages to the EMASs of the other PAs requesting hand-off and specifying the cell rates and QoS parameters of the connection(s) involved.
4. When an EMAS receives such a message it determines whether any of its PAs on the

list has sufficient capacity to accept any of the connections requiring hand-off, and responds with that information to the current EMAS.

5. When the current EMAS receives all responses back from the other EMASs, it selects a PA and communicates its choice to the mobile terminal. If the selected PA is managed by the current EMAS, we jump to step 8.

6. The current EMAS sends the new EMAS information about

   - the selected PA;
   - the list of connections to be handed off, together with their cell rates and QoS parameters;
   - the Cross Over Switch (COS), i.e. the EMAS to which the new EMAS needs to establish SVCs so as to link in to the SVCs being handed off. Figure 10.3 indicates the case where the COS is the current EMAS. In this case the current SVC is 'extended' to the new EMAS. Figure 10.4 indicates the case of 'anchor-based' re-routing in which a fixed EMAS within the current SVC is permanently designated as the COS and acts as an 'anchor' to which all other EMASs link. Another possible option for the COS is the first EMAS on the current SVC.

7. The new EMAS uses PNNI to determine a route to the COS, and then sends out M-PNNI set-up messages for each connection being handed off. When the SVCs are established, the new EMAS responds to the current EMAS that the procedure has been successful.

8. The current EMAS informs the mobile terminal that it can start communicating with the new PA. It then uses M-PNNI to release the portion of the SVCs between itself and the COS.

9. The mobile terminal executes a handshake procedure with the new EMAS in which the new EMAS informs the mobile terminal of the VPI/VCIs of each of the connections handed off.

The extensions to PNNI incorporated in M-PNNI include the ability to tear down *part* of

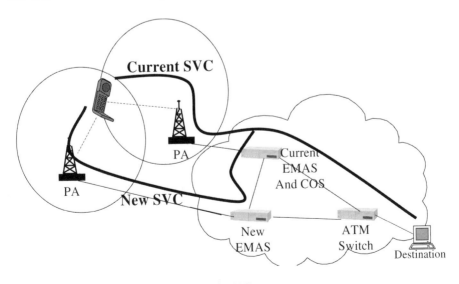

**Figure 10.3**   Extension of SVC from current EMAS

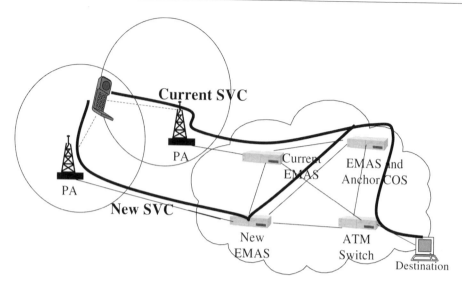

**Figure 10.4**   Anchor-based SVC re-routing

an SVC and to *add* to the remaining part of the SVC. For instance, in Figure 10.4 the part of the existing SVC from the COS to the mobile terminal via the *current* EMAS is torn down so that the part from the COS to the destination remains. The part from the COS to the mobile terminal through the *new* EMAS is added. The M-PNNI information elements required for this functionality are passed transparently through any regular ATM switches and are interpreted only by EMASs.

When an existing SVC is modified in this way it is important to maintain the QoS of the connection at a level acceptable to the user. In the case of voice, it is particularly important for the end-to-end delay to be controlled. When a connection is established using PNNI, the delay introduced by each link is cumulated in order to calculate the end-to-end delay. Referring to Figure 10.4, at the time the present SVC was set up, the COS recorded the cumulative delay, $C$, between itself and the source (the mobile terminal). It communicates this information to the current EMAS. When a hand-off occurs, the current EMAS sends the new EMAS:

- the user-requested end-to-end delay, $R$;
- the actual end-to-end delay of the current SVC, $E$;
- the cumulated end-to-end delay between the source and the COS, $C$.

When the new EMAS determines the routing from the mobile terminal to the COS, it ensures that it selects a route with end-to-end delay less than $R - (E - C)$, so that the total cumulative delay from the mobile terminal to the destination is less than $R$. A similar calculation can be performed for the case where the mobile terminal is the destination instead of the source.

## 10.2   Dense Wavelength-Division Multiplexing (DWDM)

DWDM is a method of multiplexing several colours/wavelengths of light into a single optical fibre. This dramatically increases the capacity of the optical fibre. Its impact on networking cost can be divided into three areas.

- It avoids the cost of laying additional fibre. The dispersion characteristics of some of the older installed base of fibre limit the bandwidth to 9.6 Gbps (OC-48). If 38.4 Gbps needs to be transported, OC-192 equipment cannot be used. DWDM allows 38.4 Gbps to be transported in four OC-48s, each using a separate colour of light. Even with the best quality fibre, the limit on bandwidth with a single wavelength at the time of writing is OC-768 (133.6 Gbps). If a larger amount of bandwidth needs to be transported, DWDM allows it to be done on a single fibre.
- It does not necessarily reduce the cost of the optical terminating equipment if each colour of light needs to be converted into an electrical signal. In this case a fibre carrying, for instance, 64 OC-192s on 64 different wavelengths needs 64 OC-192 terminating interfaces.
- It reduces the cost of cross connecting equipment in the network backbone. Conventional SONET/SDH cross connects convert the signal from optical to electrical and cross connect it electrically. Optical cross connects use mirrors and diffraction gratings to send each colour of the light from a given input port to the required output port.

The impact of DWDM on packet voice is two-fold.

- It reduces the cost of bandwidth and therefore reduces the need to save bandwidth by voice compression and silence removal. The evolving cost of bandwidth has to be continuously balanced against the evolving cost of voice processing chips.
- Optical cross connects cannot perform optical packet switching, but instead work on switching constant chunks of bandwidth, i.e. they do circuit switching. Much of today's packet voice implementation takes voice from an edge circuit switched network and transports it over a packet switched backbone. As optical cross connects replace packet switches in the backbone network, carriers move the packet switches to the edge, so that voice transport evolves to packet transport at the edge and circuit transport in the backbone.

## 10.3  MultiProtocol Label Switching (MPLS)

MPLS is a method of sending layer 3 protocols such as IP and Decnet over layer 2 label switching technologies such as ATM and Frame Relay. The 'label' is the VPI/VCI in ATM and the DLCI in Frame Relay. Although its name implies that MPLS is applicable to a range of layer 3 and 2 protocols, its major implementations are IP over ATM and IP over light. In the latter case the colour or wavelength, $\lambda$, of the light is used as the label, resulting in the acronym having a second meaning: MultiProtocol Lambda Switching.

MPLS is motivated by a desire to simplify the backbone of the Internet by reducing the number of entries in the input/output tables in the backbone switches. If the Internet backbone was built of high throughput IP switches/routers, they would need input/output tables indicating the correct output port for every IP destination in the world. As the number of IP addresses continuously increases, the routing tables correspondingly need to be expanded.

With MPLS, switches at the edge of the Internet backbone aggregate IP destination addresses into forwarding equivalence classes. Any IP packet needing to go to an address in a given forwarding equivalence class is sent into a Label Switched Path (LSP), which is typically an ATM virtual circuit or a wavelength of light. The LSP takes the traffic to the other side of the Internet backbone. Thus, backbone switches need input/output tables large enough to handle all current LSPs, which is far fewer than all IP destination addresses. Once the traffic exits from the LSP at the other side of the backbone, the switching is again done by examining the IP destination address. In the case of an ATM backbone, the LSP is identified by the VPI/VCIs. In the case of an optical backbone the LSP can be identified by the wavelength, $\lambda$, of the light used. An analogy can be made with the Post Office use of postal codes, or zip codes, which correspond with MPLS labels. The zip code is used by high speed letter sorting equipment in the backbone of the postal network, and the full street address (corresponding to the IP destination address) is used at the network edge. The sorting equipment *does not* need to know how to reach each street address, only how to reach each zip code. The letter carrier who delivers the letter to the addressee's house *does* need to know how to reach each street address, but only for the addresses that he/she serves, not for every address in the world.

In the case of voice traffic, MPLS is relevant for the transport of voice in IP packets across an ATM or optical backbone network. The main issue here is whether the voice traffic can be guaranteed its QoS when it is transported across multiple technologies. In the case of an optical backbone network we assume there is no problem. Nothing can get across the backbone faster than light. In the case of an ATM backbone network, it is necessary to interwork between IP QoS (RSVP and DiffServ) and ATM QoS. At the time of writing, such an interworking is in progress but has not been finalized. However, this interworking is separate from the MPLS initiative. MPLS deals with how IP destinations can be aggregated into LSPs, and how LSPs can be merged, subdivided and tunnelled through each other.

In this section, we have described MPLS very briefly. It is a major direction of network evolution, but its importance for voice traffic depends on QoS interworking which is being developed outside the MPLS forums.

## 10.4   Migration Strategies for Packet Voice

In this section we describe strategies for migrating voice traffic to packet transport for the enterprise, the local carrier and the long distance carrier.

### 10.4.1   The Enterprise

Figure 10.5 illustrates two stages of migration of an enterprise network to packet voice, from an initial architecture with circuit switched voice based on a PBX with T1/E1 access to the public carrier network. Data traffic from an ethernet LAN uses a separate access device to another T1/E1 access line.

In the first stage of evolution the access device is upgraded with a T1/E1 port incorporating voice packetization, which is connected to the PBX, so that it is now referred to as an IAD. All voice features are provided by software on the existing PBX. If the combined voice and data traffic can be accommodated in a single T1/E1 access line, the customer has saved the cost of an entire T1/E1 access. If not, they may need two T1/E1 access line ports on the IAD or alternatively they can subscribe to DSL access. These options are independent of the technology used for packet voice: IP, ATM or Frame Relay.

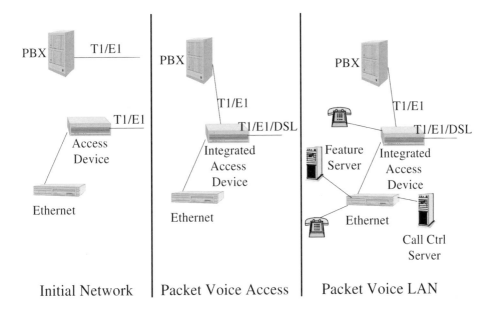

**Figure 10.5**  Enterprise network migration to packet voice

In the second stage of evolution, the voice packetization extends to the customer premises network. Although many desktop phones remain attached to the PBX and derive their features from its software, there are additional ethernet and IP phones attached to the ethernet switches and to the IAD. A call control server and a feature server are required to establish packet voice connections. Most desktop equipment is likely to use voice over IP, with Frame Relay or ATM used to transport the IP packets across either a T1/E1 or DSL access line.

### 10.4.2  The Local Carrier

We distinguish two situations:

- the ILEC *with* an installed base of class 5 edge circuit switches;
- the CLEC *without* an installed base of class 5 edge circuit switches.

Figure 10.6 illustrates the ILEC situation before and after the introduction of packet voice. Before, the class 5 edge switch provides call control and call feature functionality using the SS7 network for signalling, and the circuit switched PSTN for transport. The carrier also has an ATM and/or IP network in place for data traffic. The reason for introducing packet voice could be an increase in voice traffic beyond the capacity of the class 5 switch as a result of the increased demand for telephony and also as a result of the demand for dial-up modem access to the public Internet. The additional traffic can be migrated off the class 5 switch onto a voice gateway, which requires a wide range of customer interfaces including DSL, ISDN, SONET/SDH, T1/E1 and analogue. The call control and call feature functionality can be provided by the class 5 switch, e.g. using MEGACO (Chapter 3). Even though the class 5 switch may have exhausted its capacity for switching voice traffic, it may still have call processing capacity, or that capacity can be

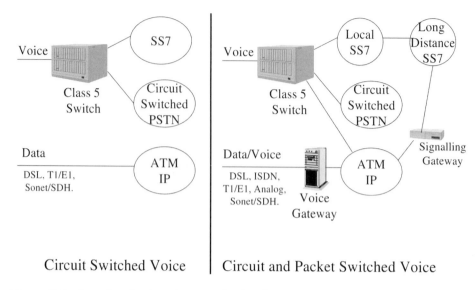

Circuit Switched Voice | Circuit and Packet Switched Voice

**Figure 10.6** Incumbent local carrier network migration to packet voice

upgraded by installing a faster processor. The calls being transported over the packet network do not need to use the SS7 network for signalling until they reach the far side of the packet voice network, where they require to interwork with SS7 through a signalling gateway *if* they need to traverse the long distance circuit switched PSTN in order to reach their destinations. Alternatively, the signalling for the packet switched voice calls can be transported over the existing local SS7 network to the long distance carrier's SS7 network.

Figure 10.7 illustrates the corresponding CLEC situation, with an initial packet data network but no circuit switched voice network. Some data comes direct from the customer over DSL, SONET/SDH and T1/E1. Other data comes from the customer via the ILEC over SONET/SDH and T1/E1. If the CLEC wishes to enter the voice business, it needs to install not only a voice gateway, but also gatekeeper, call control, feature and audio messaging servers, since it does not have an existing class 5 switch with the equivalent functionality. The voice gateway may be collocated in an ILEC CO, in which case it will require the full range of customer interfaces, including DSL, ISDN, T1/E1, analogue, and SONET/SDH. Alternatively, it may be located remotely using technologies such as ATM LES (Section 9.4) to bring traffic across the ILEC access network over T1/E1 or SONET/SDH interfaces.

### 10.4.3 The Long Distance Carrier

The two main distinctions between the long distance carrier and the ILEC/CLEC situation are:

- It is not necessary for the long distance carrier to provide voice features since they have already been provided at the edge of the network, typically by an ILEC class 5 switch, or alternatively by a call feature server.

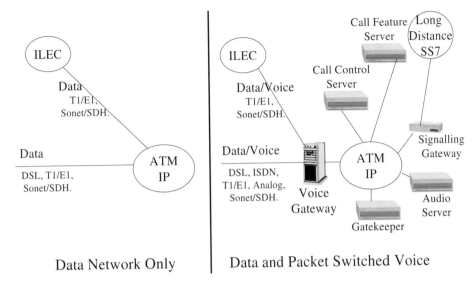

**Data Network Only** | **Data and Packet Switched Voice**

**Figure 10.7**   Competitive local carrier network migration to packet voice

- There is less need for a wide range of interfaces on the voice gateway, since it is connected to an ILEC or CLEC network, not directly to the customer.

As in the previous section, we distinguish two situations:

- the long distance carrier *with* an installed base of class 5 edge circuit switches;
- the long distance carrier *without* an installed base of class 5 edge circuit switches.

Figure 10.8 illustrates the first situation before and after the introduction of packet voice. Before, the local carrier's SS7 network is connected to the long distance carrier's toll switch and SS7 network, in order to set up calls through the toll switch, onto the long distance PSTN. The long distance carrier also has an ATM and/or IP network in place for data traffic. The reason for introducing packet voice could be an increase in voice traffic beyond the capacity of the toll switch as a result of the increased demand for telephony. The additional traffic can be migrated off the toll switch onto a voice gateway. The call establishment on the voice gateway is controlled by the toll switch via a protocol such as MEGACO. The toll switch is acting as gatekeeper, signalling gateway and call control server. Call features and audio messaging are provided in the local carrier network. The voice gateway is connected to the local carrier network via a T1/E1 or SONET/SDH interface.

Figure 10.9 illustrates the situation where the long distance carrier does not have a circuit switched network. In this case the functionality that could have been provided by the toll switch is now provided by servers and gateways. We therefore require a signalling gateway to interface to the local carrier's SS7 network plus a gatekeeper and call control server. If at the far side of the long distance carrier's network the calls need to traverse another carrier's circuit switched network, a signalling gateway is required to interface to that carrier's SS7 network.

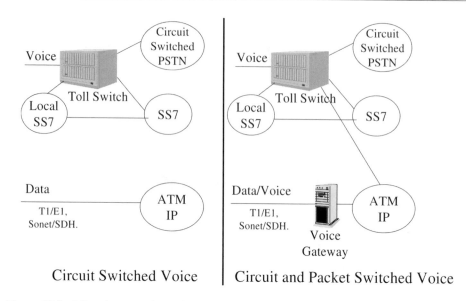

Circuit Switched Voice | Circuit and Packet Switched Voice

**Figure 10.8**   Migration to packet voice for long distance carrier *with* circuit network

### 10.4.4   Summary

The requirement for servers, gateways and interfaces in the public network scenarios discussed in Sections 10.4.2 and 10.4.3 are summarized in Table 10.1. If a function is required, this is indicated in the table by 'Yes'. If not, the alternative equipment providing the corresponding function is indicated.

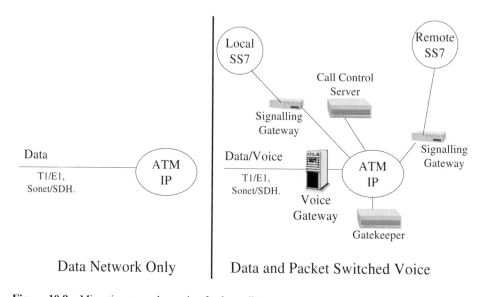

Data Network Only | Data and Packet Switched Voice

**Figure 10.9**   Migration to packet voice for long distance carrier *without* circuit network

**Table 10.1**   Location of packet voice functionality for four types of public carrier

|  | ILEC | CLEC | Long distance carrier with circuit network | Long distance carrier without circuit network |
|---|---|---|---|---|
| Call control server | Class 5 | Yes | Toll switch | Yes |
| Call feature server | Class 5 | Yes | ILEC class 5 | ILEC class 5 |
| Gatekeeper | Class 5 | Yes | Toll switch | Yes |
| Audio server | Class 5 | Yes | ILEC class 5 | ILEC class 5 |
| Signalling gateway | Class 5 | No | Toll switch | Yes |
| Voice gateway | Yes | Yes | Yes | Yes |
| T1/E1, SONET/SDH | Yes | Yes | Yes | Yes |
| DSL, ISDN, analogue | Yes | Yes | ILEC class 5 | ILEC class 5 |

## Reference

1. ATM Forum (2000) Wireless ATM Capability Set Specification.

# Epilogue

## Voice: Still Searching for Its Holy Grail

The winds that blow voice from circuit to packet transport also whistle eerily though the empty production facilities that once rolled out dinosaur-style circuit switches. They fan the flames that burn the heels of those telephone companies that lagged in the race to migrate voice away from the circuit networks that they had built and fine tuned over decades. And they cause waves that dash upon rocky shores of the careers of men and women who did not see the direction in which the weather vane was pointing.

Winds of change that cause havoc for those that cling to former ways, bring opportunity to those who fly with them. The flags fly high and strong over the packet switch lines of business in many equipment vendors, large and small. Those who work in silicon, embrace the need for voice codecs, echo cancellers and more on one chip. And those who deploy packet voice technology rejoice in the low maintenance costs that hardware implementation brings about. The government regulators, in their ivory towers, celebrate how packet transport has helped them stimulate that same competitive spirit in the telecommunications industry, that is such an anathema to their own work environment. And last, but not least, are the customers, talking for hours to distant relatives at tariffs lower than they ever imagined a decade ago.

The winds that blow voice today, stream unswervingly past weather vanes pointing them clearly away from circuit and towards packet. But will they one day change direction? Is voice transport more like a sail of a windmill, eventually coming full circle? Could voice transport eventually return to circuit technology? If we forget the regulatory environment that has effectively subsidized voice over IP, and forget the age of the circuit equipment that makes it artificially more costly than modern packet switches; then packet transport does one thing and one thing only: it saves bandwidth during voice silence periods. And winds of change in the telecommunications industry are multidimensional. They blow voice from circuit to packet, and at the same time enable hundreds of OC768s to be transported on a single fiber. Question: "Why save bandwidth, if bandwidth is cheap?". Answer: "If saving bandwidth is cheaper." Everything is relative. If the silicon that packetizes the voice costs less than the bandwidth that is saved, we will still packetize voice, even when bandwidth is cheap. But if silicon costs more than bandwidth, voice will go full circle, back to circuit transport.

No one who has participated in the telecommunications industry over the last few decades has found a killer technology. ISDN, ATM, IP, DWDM: these are all leaves blowing in the wind. Will voice find its holy grail in one of these? Not for my money. There will always be another technology on the horizon, waiting to ensnare the naive with claims of omnipotence, waiting to destroy those who cling to the past, and waiting to create new opportunities for those who critically analyze its potentials.

May I wish all my readers that the next decade of their professional lives be as exciting and productive as the past decade has been for me.

# Acronyms

| | |
|---|---|
| **AAL** | ATM Adaptation Layer |
| **AAL1** | ATM Adaptation Layer Type 1 |
| **AAL2** | ATM Adaptation Layer Type 2 |
| **AAL5** | ATM Adaptation Layer Type 5 |
| **ABR** | Available Bit Rate |
| **ADPCM** | Adaptive Differential Pulse Code Modulation |
| **ADSL** | Asymmetric Digital Subscriber Line |
| **ADT** | Assured Data Transfer |
| **AF** | Assured Forwarding. A Diffserv Per Hop Behaviour |
| **AIS** | Alarm Indication Signal |
| **ANSI** | American National Standards Institute |
| **API** | Applications Programming Interface |
| **ATM** | Asynchronous Transfer Mode |
| **ATU-C** | ADSL Terminal Unit – Central Office |
| **ATU-R** | ADSL Terminal Unit – Remote |
| **AVP** | Audio Video Protocol |
| **B$_c$** | Committed Burst |
| **B$_e$** | Excess Burst |
| **BECN** | Backward Explicit Congestion Notification |
| **BER** | Bit Error Rate |
| **B-ISDN** | Broadband ISDN |
| **BLES** | Broadband Loop Emulation Service |
| **BRI** | Basic Rate Interface |
| **CAS** | Channel Associated Signalling |
| **CBR** | Constant Bit Rate |
| **CCS** | Common Channel Signalling |
| **CDR** | Call Detail Record |
| **CDV** | Cell Delay Variation |
| **CES** | Circuit Emulation Service |
| **CI** | Congestion Indication |
| **CID** | Channel Identifier |
| **CIR** | Committed Information Rate |
| **CLASS** | Customer Local Area Signalling Service |
| **CLE** | Customer Located Equipment |
| **CLEC** | Competitive Local Exchange Carrier |
| **CLP** | Cell Loss Priority |
| **CLR** | Cell Loss Ratio |
| **CM** | Cable Modem |
| **CMIP** | Common Management Information Protocol |

| **CMS** | Call Management Server |
|---|---|
| **CMTS** | Cable Modem Termination System |
| **CO** | Central Office |
| **Codec** | COder-DECoder |
| **CoS** | Class of Service |
| **COS** | Cross Over Switch |
| **CPS** | Common Part Sublayer |
| **CRC** | Cyclical Redundancy Check |
| **CSI** | Convergence Sublayer Indication |
| **CTD** | Cell Transfer Delay |
| **DBCES** | Dynamic Bandwidth Circuit Emulation Service |
| **DE** | Discard Eligibility |
| **DiffServ** | Differentiated Services |
| **DLCI** | Data Link Connection Identifier |
| **DMT** | Discrete MultiTone |
| **DNS** | Domain Name Server |
| **DOCSIS** | Data Over Cable System Interface Specification |
| **DPCM** | Differential Pulse Code Modulation |
| **DSCP** | Diffserv Code Point |
| **DSL** | Digital Subscriber Line |
| **DSLAM** | Digital Subscriber Line Access Multiplexer |
| **DSP** | Digital Signal Processor |
| **DTE** | Data Terminal Equipment |
| **DTL** | Designated Transit List |
| **DTMF** | Dual Tone Multi-Frequency |
| **DWDM** | Dense Wavelength Division Multiplexing |
| **EA** | Extended Address |
| **EF** | Expedited Forwarding |
| **EI** | Extension Indicator |
| **EMAS** | End-User Mobility ATM Switch |
| **ETSI** | European Telecommunications Standards Institute |
| **FCC** | Federal Communications Commission |
| **FCS** | Frame Check Sequence |
| **FDDI** | Fiber Distributed Data Interface |
| **FEC** | Forward Error Correction |
| **FECN** | Forward Explicit Congestion Notification |
| **FRAD** | Frame Relay Access Device |
| **GC** | Gate Controller |
| **GFC** | Generic Flow Control |
| **GFR** | Guaranteed Frame Rate |
| **GPS** | Global Positioning Satellite |
| **GSM** | Global System for Mobile Communications |
| **HDLC** | High-Level Data Link Control |
| **HDSL** | High Speed Digital Subscriber Line |
| **HEC** | Header Error Control |
| **HFC** | Hybrid Fibre/Coax |
| **HTTP** | Hyper Text Transfer Protocol |

| | |
|---|---|
| **IA** | Implementation Agreement |
| **IAD** | Integrated Access Device |
| **IANA** | Internet Assigned Numbers Authority |
| **IEC** | Inter Exchange Carrier |
| **IETF** | Internet Engineering Task Force |
| **ILEC** | Incumbent Local Exchange Carrier |
| **IP** | Internet Protocol |
| **ISDN** | Integrated Services Digital Network |
| **ISP** | Internet Service Provider |
| **ITSP** | Internet Telephony Service Provider |
| **ITU** | International Telecommunication Union |
| **ITU-T** | International Telecommunication Union – Telecommunication Standardization Sector |
| **ITXC** | Internet Telephony Exchange Carrier |
| **IVR** | Interactive Voice Response |
| **IWF** | InterWorking Function |
| **IXC** | Inter Exchange Carrier |
| **LAN** | Local Area Network |
| **LEC** | Local Exchange Carrier |
| **LEO** | Low Earth Orbit |
| **LES** | Loop Emulation Service |
| **LI** | Length Indicator |
| **LNP** | Local Number Portability |
| **LoF** | Loss of Frame |
| **LoS** | Loss of Signal |
| **LP** | Loss Priority |
| **LPC** | Linear Predictive Coding |
| **LSB** | Least Significant Bit |
| **LSP** | Label Switched Path |
| **MBS** | Maximum Burst Size |
| **MCU** | Multipoint Control Unit |
| **MEGACO** | Media Gateway Control |
| **MEO** | Medium Earth Orbit |
| **MG** | Media Gateway |
| **MGC** | Media Gateway Controller |
| **MGCP** | Media Gateway Control Protocol |
| **MIB** | Management Information Base |
| **MIME** | Multimedia Internet Mail Extension |
| **MIPS** | Mega Instructions Per Second |
| **MOS** | Mean Opinion Score |
| **MPEG** | Motion Picture Experts Group |
| **MPLS** | MultiProtocol Label Switching |
| **M-PNNI** | Mobile Private Network-to-Network Interface |
| **MRI** | Magnetic Resonance Imaging |
| **MSB** | Most Significant Bit |
| **MTA** | Media Terminal Adapter |
| **MTU** | Maximum Transmission Unit |

| | |
|---|---|
| **M-UNI** | Mobile User-to-Network Interface |
| **NEBS** | National Environmental Building Standards |
| **NLPID** | Network Layer Protocol Identifier |
| **NNI** | Network-to-Network Interface |
| **nrt-VBR** | Non-Real Time Variable Bit Rate |
| **NSAP** | Network Service Access Point |
| **OAM** | Operation Administration and Maintenance |
| **OSI** | Open Systems Interconnection |
| **OSPF** | Open Shortest Path First |
| **PA** | Point of Attachment |
| **PAMS** | Perceptual Analysis Measurement System |
| **PBX** | Private Branch Exchange |
| **PCM** | Pulse Code Modulation |
| **PDU** | Protocol Data Unit |
| **PNNI** | Private Network-to-Network Interface |
| **POTS** | Plain Old Telephone Service |
| **PPP** | Point-to-Point Protocol |
| **PRI** | Primary Rate Interface |
| **PSQM** | Perceptual Speech Quality Measurement |
| **PSTN** | Public Switched Telephone Network |
| **PT** | Payload Type |
| **PTI** | Payload Type Indicator |
| **PVC** | Permanent Virtual Circuit |
| **QoS** | Quality of Service |
| **RAI** | Remote Alarm Indication |
| **RDI** | Remote Defect Indication |
| **RKS** | Record Keeping Server |
| **RSVP** | ReSource reserVation Protocol |
| **RTCP** | Real Time Control Protocol |
| **RTP** | Real Time Protocol |
| **rt-VBR** | Real Time Variable Bit Rate |
| **SAAL** | Signalling ATM Adaptation Layer |
| **SAR** | Segmentation and Reassembly |
| **SCP** | Service Control Point |
| **SCR** | Sustainable Cell Rate |
| **SDH** | Synchronous Digital Hierarchy |
| **SDP** | Session Description Protocol |
| **SDSL** | Symmetrical Digital Subscriber Line |
| **SDT** | Structured Data Transfer |
| **SG** | Signalling Gateway |
| **SGCP** | Simple Gateway Control Protocol |
| **SID** | Silence Insertion Descriptor |
| **SIP** | Session Initiation Protocol |
| **SLA** | Service Level Agreement |
| **SME** | Small/Medium Enterprise |
| **SN** | Sequence Number |
| **SNMP** | Simple Network Management Protocol |

| | |
|---|---|
| **SOHO** | Small Office/Home Office |
| **SONET** | Synchronous Optical NETwork |
| **SRTS** | Synchronous Residual Time Stamp |
| **SS7** | Signalling System No. 7 |
| **SSADT** | Service-Specific Assured Data Transfer |
| **SSCOP** | Service-Specific Connection Oriented Protocol |
| **SSCS** | Service-Specific Convergence Sublayer |
| **SSSAR** | Service-Specific Segmentation and Reassembly |
| **SSTED** | Service-Specific Transmission Error Detection |
| **STM** | Synchronous Transport Module |
| **STS** | Synchronous Transport Signal |
| **SVC** | Switched Virtual Circuit |
| **TAPI** | Telephony Applications Programming Interface |
| **TCP** | Transmission Control Protocol |
| **TDM** | Time Division Multiplexed |
| **TIFF** | Tag Image File Format |
| **ToS** | Type of Service |
| **TSAPI** | Telephony Services Applications Programming Interface |
| **UAC** | User Agent Client |
| **UAS** | User Agent Server |
| **UBR** | Unspecified Bit Rate |
| **UDP** | User Datagram Protocol |
| **UDT** | Unstructured Data Transfer |
| **UNI** | User-to-Network Interface |
| **USB** | Universal Serial Bus |
| **UUI** | User-to-User Information |
| **VAD** | Voice Activity Detection |
| **VBR** | Variable Bit Rate |
| **VC** | Virtual Circuit |
| **VCC** | Virtual Channel Connection |
| **VCI** | Virtual Channel Identifier |
| **VDSL** | Very High Speed Digital Subscriber Line |
| **Vocoder** | Voice coder/decoder |
| **VPI** | Virtual Path Identifier |
| **VPIM** | Voice Profile for Internet Mail |
| **WAN** | Wide Area Network |
| **WATM** | Wireless ATM |

# Standards Bibliography

This bibliography contains references to standards, documents and implementation agreements related to packet voice. It is arranged according to the organization producing the document in the following sequence: ITU-T, IETF, ATM Forum, Frame Relay Forum, CableLabs, DSL Forum.

## ITU-T: International Telecommunication Union: Standards

http://www.itu.ch/

| | |
|---|---|
| E.180/Q.35 | Recommendation E.180/Q.35 (1998) Technical Characteristics of Tones for the Telephone Service. |
| G.126 | Recommendation G.126 (1993) Listener Echo in Telephone Networks. |
| G.131 | Recommendation G.131 (1996) Control of Talker Echo. |
| G.164 | Recommendation G.164 (1988) Echo Suppressors. |
| G.165 | Recommendation G.165 (1993) Echo Cancellers. |
| G.167 | Recommendation G.167 (1993) Acoustic Echo Controllers. |
| G.168 | Recommendation G.168 (1997) Digital Network Echo Cancellers. |
| G.711 | Recommendation G.711 (1988) Pulse Code Modulation (PCM) of Voice Frequencies. |
| G.722 | Recommendation G.722 (1988) 7 kHz Audio-Coding Within 64 kbit/s. |
| G.723.1 | Recommendation G.723.1 (1996) Dual Rate Speech Coder for Multimedia Communications Transmitting at 5.3 and 6.3 kbit/s. |
| G.726 | Recommendation G.726 (1990) 40, 32, 24, 16 kbit/s Adaptive Differential Pulse Code Modulation (ADPCM). |
| G.727 | Recommendation G.727 (1990) 5-, 4-, 3- and 2-Bits Sample Embedded Adaptive Differential Pulse Code Modulation (ADPCM). |
| G.728 | Recommendation G.728 (1992) Coding of Speech at 16 kbit/s Using Low-Delay Code Excited Linear Prediction. |
| G.728H | Recommendation G.728 Annex H (1997) Variable Bit Rate LD-CELP Operation Mainly for DCME at Rates Less Than 16 kbit/s. |
| G.729 | Recommendation G.729 (1996) Coding of Speech at 8 kbit/s Using Conjugate Structure Algebraic-Code-Excited Linear-Prediction (CS-ACELP). |
| G.729A | Recommendation G.729 Annex A (1996) Reduced Complexity 8 kbit/s CS-ACELP Speech Coder. |
| G.729B | Recommendation G.729 Annex B (1996) A Silence Compression Scheme for G.729 Optimized for Terminals Conforming to Recommendation V.70. |
| G.729D | Recommendation G.729 Annex D (1998) 6.4 kbit/s CS-ACELP Speech Coding Algorithm. |
| G.729E | Recommendation G.729 Annex E (1998) 11.8 kbit/s CS-ACELP Speech Coding Algorithm. |
| G.992.1 | Asymmetrical Digital Subscriber Line (ADSL) Transceivers. |
| G.992.2 | Splitterless Asymmetrical Digital Subscriber Line (ADSL) Transceivers. |

| H.221 | Recommendation H.221 (1999) Frame Structure for a 64 to 1920 kbit/s Channel in Audiovisual Teleservices. |
|---|---|
| H.223 | Recommendation H.223 (1996) Multiplexing Protocol for Low Bit Rate Multimedia Communication. |
| H.225.0 | Recommendation H.225.0 (1998) Call Signaling Protocols and Media Stream Packetization for Packet Based Multimedia Communications Systems. |
| H.235 | Recommendation H.235 (1998) Security and Encryption for H-Series (H.323 and Other H.245-Based) Multimedia Terminals. |
| H.245 | Recommendation H.245 (1998) Control Protocol for Multimedia Communication. |
| H.246 | Recommendation H.246 (1998) Interworking of H-Series Multimedia Terminals With H-Series Multimedia Terminals and Voice/Voiceband Terminals on GSTN and ISDN. |
| H.248 | Recommendation H.248 (2000) Gateway Control Protocol. |
| H.261 | Recommendation H.261 (1993) Video Codec for Audiovisual Services at $p \times 64$ kbit/s. |
| H.263 | Recommendation H.263 (1996) Video Coding for Low Bit Rate Communication. |
| H.320 | Recommendation H.320 (1997) Narrow-Band Visual Telephone Systems and Terminal Equipment. |
| H.321 | Recommendation H.321 (1996) Adaptation of H.320 Visual Telephone Terminals to B-ISDN Environments. |
| H.322 | Recommendation H.322 (1996) Visual Telephone Systems and Terminal Equipment for Local Area Networks Which Provide a Guaranteed Quality of Service. |
| H.323 | Recommendation H.323 (1999) Packet-Based Multimedia Communications Systems, January 1998. |
| H.324 | Recommendation H.324 (1996) Terminal for Low Bit Rate Multimedia Communication. |
| I.363.1 | Recommendation I.363.1 (1996) B-ISDN ATM Adaptation Layer Specification: Type 1 AAL. |
| I.363.2 | Recommendation I.363.2 (1997) B-ISDN ATM Adaptation Layer Type 2 Specification. |
| I.363.5 | Recommendation I.363.5 (1996) B-ISDN ATM Adaptation Layer Specification: Type 5 AAL. |
| I.366.1 | Recommendation I.366.1 (1998) Segmentation and Reassembly Service Specific Convergence Sublayer for the AAL Type 2. |
| I.366.2 | Recommendation I.366.2 (1999) AAL Type 2 Service Specific Convergence Sublayer For Trunking. |
| I.371 | Recommendation I.371 (1996) Traffic Control and Congestion Control in B-ISDN. |
| P.861 | Recommendation P.861 (1998) Objective Quality Measurement of Telephone-Band (300–3400 Hz) Speech Codecs. |
| Q.933 | Recommendation Q.933 (1995) Signalling Specifications for Frame Mode Switched and Permanent Virtual Connection Control and Status Monitoring. |
| Q.2630.1 | Recommendation Q.2630.1 (1999) AAL Type 2 Signalling Protocol (Capability Set 1). |
| Q.2931 | Recommendation Q.2931 (1995) User-Network Interface – Layer 3 Specification for Basic Call Connection Control. |
| T.37 | Recommendation T.37 (1998) Procedures for the Transfer of Facsimile Data Via Store-and-Forward on the Internet. |
| T.120 | Recommendation T.120 (1996) Data Protocols for Multimedia Conferencing. |
| V.17 | Recommendation V.17 (1991) A 2-Wire Modem for Facsimile Applications With Rates up to 14400 bit/s. |
| V.21 | Recommendation V.21 (1988) 300 Bits Per Second Duplex Modem Standardized for Use in the General Switched Telephone Network. |
| V.27ter | Recommendation V.27ter (1988) 4800/2400 Bits Per Second Modem Standardized for Use in the General Switched Telephone Network. |

V.29          Recommendation V.29 (1988) 9600 Bits Per Second Modem Standardized for Use on
              Point-to-Point 4-Wire Leased Telephone-Type Circuits.

## IETF: Internet Engineering Task Force. RFCs, Requests for Comments

www.ietf.org

RFC 768      Postel, J. (1980) *User Datagram Protocol.*
RFC 791      Postel, J. (1981) *Internet Protocol.*
RFC 793      Postel, J. (1981) *Transmission Control Protocol.*
RFC 1483     Heinanen, J. (1993) *Multiprotocol Encapsulation Over ATM Adaptation Layer 5.*
RFC 1577     Laubach, M. (1994) *Classical IP and ARP Over ATM.*
RFC 1889     Schulzrinne, H., Casner, S., Frederick, R. and Jacobson, V. (1996) *RTP: A Transport
             Protocol for Real-Time Applications. Audio-VideoTransport Working Group.*
RFC 1890     Schulzrinne, H. (1996) *RTP Profile for Audio and Video Conferences With Minimal
             Control. Audio-Video Transport Working Group.*
RFC 2205     Braden, R., et al. (1997) *Resource ReSerVation Protocol (RSVP) – Version 1 Func-
             tional Specification.*
RFC 2209     Braden, R., Zhang, L., Berson, S., Herzog, S. and Jamin, S. (1997) *Resource ReSerVa-
             tion Protocol (RSVP) – Version 1 Functional Specification.*
RFC 2210     Wroclawski, J. (1997) *The Use of RSVP With IETF Integrated Services.*
RFC 2212     Shenker, S. and Partridge, C. (1997) *Specification of Guaranteed Quality of Service.*
RFC 2301     McIntyre, L., Zilles, S., Buckley, R., Venable, D., Parsons, G. and Rafferty, J. (1998)
             *File Format for Internet Fax.*
RFC 2302     Parsons, G., Rafferty, J. and Zilles, S. (1998) *Tag Image File Format (TIFF) – image/
             tiff MIME Sub-Type Registration.*
RFC 2306     Parsons, G. and Rafferty, J. (1998) *Tag Image File Format (TIFF) – F Profile for
             Facsimile.*
RFC 2326     Schulzrinne, H., Rao, A. and Lanphier, R. (1998) *Real Time Streaming Protocol
             (RTSP).*
RFC 2327     Handley, M. and Jacobson, V. (1998) *SDP: Session Description Protocol.*
RFC 2382     Crawley, E., Berger, L., Berson, S., Baker, F., Borden, M. and Krawczyk, J. (1998) *A
             Framework for Integrated Services and RSVP Over ATM.*
RFC 2386     Rajagopalan, B., Nair, R., Sandick, H. and Crawley, E. (1998) *A Framework for QoS-
             Based Routing in the Internet.*
RFC 2421     Vaudreuil, G. and Parsons, G. (1998) *VPIM, Voice Profile for Internet Mail.*
RFC 2460     Deering, S. and Hinden, R. (1998) *Internet Protocol, Version 6 (IPv6) Specification.*
RFC 2508     Casner, S. and Jacobson, V. (1999) *Compressing IP/UDP/RTP Headers for Low-Speed
             Serial Links.*
RFC 2543     Handley, M., Schulzrinne, H., Schooler, E. and Rosenberg, J. (1999) *SIP: Session
             Initiation Protocol.*
RFC 2658     McKay, K. (1999) *RTP Payload Format for PureVoice™ Audio.*
RFC 2705     Arango, M., Dugan, A., Elliott, I., Huitema, C. and Pickett, S. (1999) *Media Gateway
             Control Protocol (MGCP) Version 1.0.*
RFC 2733     Rosenberg, J. and Schulzrinne, H. (1999) *An RTP Payload Format for Generic
             Forward Error Correction.*
RFC 2805     Greene, N., Ramalho, M. and Rosen, B. (2000) *Media Gateway Control Protocol
             Architecture and Requirements.*

RFC 2833    Schulzrinne, H. and Petrack, S. (2000) *RTP Payload for DTMF Digits, Telephony Tones and Telephony Signals.*
RFC 2885    Cuervo, F., Greene, N., Huitema, C., Rayhan, A., Rosen, B. and Segers, J. (2000) *Megaco Protocol.*
RFC 2990    Huston, J. (2000) *Next Steps for the IP QoS Architecture.*

## ATM Forum: Approved Specifications

http://www.atmforum.com

Circuit Emulation Service 2.0, af-vtoa-0078.000, January 1997.
Voice and Telephony Over ATM to the Desktop, af-vtoa-0083.001, February 1999.
(DBCES) Dynamic Bandwidth Utilization in 64 KBPS Time Slot Trunking Over ATM – Using CES, af-vtoa-0085.000, July 1997.
ATM Trunking Using AAL1 for Narrow Band Services v1.0, af-vtoa-0089.000, July 1997.
ATM Trunking Using AAL2 for Narrow Band Services, af-vtoa-0113.000, February 1999.
Low Speed Circuit Emulation Service, af-vtoa-0119.000, May 1999.
ICS for ATM Trunking Using AAL2 for Narrowband Services, af-vtoa-0120.000, May 1999.
Low Speed Circuit Emulation Service (LSCES), Implementation Conformance Statement Proforma, af-vtoa-0132.000, October 1999.
Voice and Multimedia Over ATM, Local Loop Emulation Using AAL2, af-vmoa-0145.000, July 2000.

## Frame Relay Forum: Implementation Agreements

www.frforum.com

FRF.1.1     D. Sinicrope (Ed.), User-to-Network Implementation Agreement (UNI), January 1996.
FRF.2.1     L. Greenstein (Ed.), Frame Relay Network-to-Network Interface Implementation Agreement, July 1995.
FRF.3.1     R. Cherukuri (Ed.), Multiprotocol Encapsulation Implementation Agreement, June 1995.
FRF.5       D. O'Leary (Ed.), Frame Relay/ATM PVC Network Interworking Implementation Agreement, December 1994.
FRF.8       D. O'Leary (Ed.), Frame Relay/ATM PVC Service Interworking Implementation Agreement, April 1995.
FRF.11      K. Rehbehn, R. Kocen and T. Hatala (Eds.), Voice Over Frame Relay Implementation Agreement, March 1997.
FRF.12      A.G. Malis (Ed.), Frame Relay Fragmentation Implementation Agreement, December 1997.
FRF.13      K. Rehbehn (Ed.), Service Level Definitions Implementation Agreement, August 1998.

## CableLabs Specifications

www.cablelabs.com/

Data-Over-Cable Service Interface Specifications, Radio Frequency Interface Specification, SP-RFIv1.1-I03-991105, Cable Television Laboratories, Inc., November 1999.

PacketCable Distributed Call Signaling Specification, PKT-SP-DCS-D02-991007, Cable Television Laboratories, Inc., October 1999.

PacketCable Network-Based Call Signaling Protocol Specification, PKT-SP-EC-MGCP-I02-991201, Cable Television Laboratories, Inc., December 1999.

PacketCable Audio/Video Codecs Specification, PKT-SP-CODEC-I01-991201, Cable Television Laboratories, Inc., December 1999.

PacketCable Dynamic QoS Specification, PKT-SP-DQOS-102-000818, Cable Television Laboratories, Inc., August 2000.

## DSL Forum: Approved Technical Reports

http://www.ads1.com/tr_table.html

| TR-001 | DSL Forum System Reference Model. |
|--------|-----------------------------------|
| TR-007 | Interfaces and System Configurations for ADSL: CPE. |
| TR-008 | Default VPI/VCI Addresses for FUNI Mode Transport. |
| TR-010 | Requirements/Reference Models for ADSL Access Networks. |
| TR-012 | Broadband Service Architecture (PPP over ATM). |
| TR-013 | Interface & Configurations for ADSL: Central Office. |
| TR-017 | ATM Over ADSL Recommendation. |
| TR-019 | Physical Layer of ADSLs With a Splitter. |
| TR-020 | Physical Layer of ADSLs Without a Splitter. |
| TR-021 | ADSL Forum Recommendation for ATM Layer of ADSLs. |
| TR-036 | Requirements for Voice Over DSL. |

# Index